The E-Commerce Book
Building the E-Empire

The E-Commerce Book
Building the E-Empire

Steffano Korper

Co-founder of the E-Commerce Program
Dallas, Texas

Juanita Ellis

Co-founder of the E-Commerce Program
Los Angeles, California

ACADEMIC PRESS

A Harcourt Science and Technology Company

San Diego San Francisco New York Boston
London Sydney Tokyo

Cover photo by:
Pearle Productions
1117 N. Canterbury Court
Dallas, Texas 75208
www.pearleproductions.com

ACADEMIC PRESS
A Harcourt Science and Technology Company
525 B Street, Suite 1900, San Diego, CA 92101-4495, USA
http://www.apnet.com

Academic Press
24–28 Oval Road, London NW1 7DX, UK
http://www.hbuk.co.uk/ap/

Library of Congress Catalog Card Number: 99-64978

International Standard Book Number: 0-12-421160-7

Printed in the United States of America
99 00 01 02 03 IP 9 8 7 6 5 4 3 2 1

Contents

Foreword

Only a few short years ago, a chaired professor of engineering at a major university remarked in passing, "The Internet has no commercial viability." The statement may have been true at that time, but even then it was clear to many others that the Internet provided access, and access is the core of commerce. If you picked up this book, you suspect e-commerce is something you need. Now you want to know: What is it? How do I get it? How do I manage it? The answers are in this book.

What is it? From the business-to-business viewpoint, it is the next "best practices." From the business-to-consumer viewpoint, it is the ultimate retail opportunity. From the individual's viewpoint, it is every person's connection to the global economy. But, from all viewpoints, it is the flow of money, moving quickly and silently to those who are sufficiently wise and creative to establish themselves as players in this new arena.

Steffano Korper and Juanita Ellis are experts in e-commerce, and there are precious few experts today. They have broad experience in information technology, software, business, and computing, and both consult regularly with a host of corporations on e-commerce needs. They are also extraordinary communicators and energetic instructors of e-commerce courses. Moreover, they are innovators and visionaries, and they can help you envision and realize your opportunity in e-commerce.

This volume, their first book on e-commerce, is a quick read and is loaded with content. An unusual combination, indeed! The book is written for a broad audience, certainly NOT just for technologists. As you read, the opportunities provided by e-commerce will open up to you. And you will feel the incredible energy available by tapping directly into the global economy of e-commerce.

If you want to see the bottom line first, I recommend reading the Project Deployment chapter immediately. This chapter outlines the entire process of getting your e-commerce solution in place. With the knowledge and motivation from this chapter, go back and read the rest of the book to fill in the details.

You can be part of the electronic revolution. It's happening now. This is your opportunity.

Jerry D. Gibson, Chair
Department of Electrical Engineering
School of Engineering and Applied Science
Southern Methodist University
Dallas, Texas

Preface

A year ago, we read the first financial estimate for electronic commerce. "Internet commerce is transforming the way enterprises do business," one report stated in May, 1998. "By 2002, the value of goods and services traded between companies over the Internet will represent 1 percent of the global economy (approximately $317 billion)."

Now, just a year later, that amazing figure is a gross underestimate. Electronic commerce (or e-commerce) is the buying and selling of goods and services over the Internet, and it is rapidly changing the way successful people conduct business. In fact, forward-thinking individuals see e-commerce as the next wave of success for all businesses, from those serving three customers a week to those who serve three customers each second. E-commerce is the next Industrial Revolution because of its potential to affect fundamental changes in the way people and companies buy and sell products and services.

Respected research and development firms support such predictions of a radical new marketplace without physical or fiscal limits. In January 1999, Forrester Research increased its earlier estimate for Internet trade four-fold to $1.3 trillion. That $1.3 trillion figure was stale by early summer, when International Data Corp. estimated 1999 expenditures on the Internet would total $479 billion worldwide, including $174 billion in the United States alone. That same report forecasted corporate e-commerce expenditures would reach $2.2 trillion by 2003. These estimates imply that businesspeople's window of opportunity, the potential for earnings on any e-commerce investment, lies somewhere between $479 billion and $2.2 trillion in the next four years.

Now that the World Wide Web's fiscal promise is certain, the only question from future-minded businesspeople should be how do I immediately get my company on board and maximize its earnings? A few companies, including Amazon.com and eBay, developed an answer early and were rewarded with household name recognition and soaring stock prices. Luckily for everyone else, the Internet's extraordinary youth means its earnings potential has no apparent ceiling. Plenty of room still exists for pioneers to enjoy similar e-commerce success. This book will show them how to achieve it.

The E-Commerce Book: Building the E-Empire is a direct result of more than ten years of industry experience and education. Its purpose is to explain how any company or individual can help build the E-Empire and benefit from an expected total windfall of $2.2 trillion in 2003. Its goal is to give each reader the right tools to jump head-first into the pool of e-commerce and to find it comfortable and deep with opportunity.

The following chapters provide the secrets that all businesspeople, including small entrepreneurs, corporate gurus, and managers on every level, need to make their business succeed in an e-commerce venture. Included are directions for building the correct infrastructure for an efficient and profitable e-commerce venture, and instructions and advice on using that infrastructure to create virtual storefronts on Web sites, to fashion those storefronts into online shopping malls, and to send targeted e-mails and faxes that will encourage new markets and support existing ones. The book also shares with each reader the secrets of integrating electronic data interchange (EDI) into existing business structures, of conducting business-to-business commerce, of ensuring secure and convenient transactions, and of discovering compelling processes that separate a successful e-commerce company from its competition.

The book contains new, cutting-edge information that may intimidate those who are unfamiliar with the Internet or are reluctant to explore the new, potentially lucrative market it sustains. Those who proceed, read the book, and use it as a tool to expand their companies' markets will be the first shareholders in the newest pairing of global business and top-notch technology...the E-Empire.

Our vision is to provide the tools and knowledge you need to be a "Mover and Shaker" in the e-commerce marketplace. This is the first book in a series that we intend to share with our readers. As this evolution continues, we will keep you updated on what you need to know in the e-commerce arena and how to implement the knowledge you have acquired. We want this book to generate the same enthusiasm and knowledge, for everyone involved in the business, as we have experienced with so many professionals and corporations thoughout the United States. Whether you are the president of a company, the owner of a small business, or an entrepreneur working from home...ENJOY THE RIDE!

The E-Commerce Book
Building the E-Empire

1

E-Commerce Overview

- How companies have been successful in e-commerce
- E-commerce business models
- Gaining the competitive advantage
- Redefining the business process

The past few years have ushered in enormous changes in the way companies are doing business, selling goods and services, and communicating with their suppliers and buyers. The large brick-and-mortar companies are rethinking their business models in order to compete in the new marketplace. Small electronic communities enter the market with innovative ways to sell their goods and services. This is a time of remarkable opportunity for those businesses that harness the power of the new market, namely electronic commerce. Those companies that underestimate its power may be left behind, as other companies flourish in the new environment.

Electronic commerce will play a major role in the way small, medium, and large companies conduct business either with their consumers, other businesses or both. It is critical to understand the e-commerce market early in the game and to understand how e-commerce changes your business model. Now is the time to reconsider the way you are doing business and how you should approach the new global electronic community. Your competitor is thinking about this very thing.

Now is also the time for entrepreneurs and smaller businesses to compete on the same playing field as larger corporations. In the electronic community, a lack of real estate is not an obstacle. A lack of vast resources, such as employees and capital, is not an obstacle. Your ideas, innovation, and drive carry your ideas forward. Most of the top-selling e-commerce companies are less than five years old. Your focus remains on generating the right idea at the right time. Think of a time that you did something right. What was the result? What was the reward? The inherent beauty of electronic commerce is that when you do one thing right, you get paid over and over.

The objective of this chapter is to provide you with the foundation required to understand e-commerce and how your business fits into the current marketplace. Emphasis is placed on business models, direct selling models, and the key factors required to successfully deploy an e-commerce solution. As you read though this chapter, consider your current business model and how electronically enabling your enterprise, of whatever size, provides direct benefits to your customers and business partners.

Technology Growth Rates

Many individuals and business leaders may be thinking that "the Internet won't change my business." Consider the growth of technology illustrated in Figure 1.1. With the Internet's explosive growth, doubting individuals may want to think again.

We have just witnessed the birth of e-commerce. Now, consider its rate of growth. Consider the not too distant future. People are accessing the Internet today primarily with a desktop computer. In the near future, that will change, especially in the consumer market. With products like Web TV becoming increasingly popular, consumers are able to sit in their living rooms and shop in three-dimensional malls, make travel plans by viewing resorts and accommodations online, and order educational videos on the topic of their choice. Of course, we will continue to use the computer to access the Internet. However, with Web TV and other less traditional means of Internet access, your business can capture audiences that generally have not grown up with computers and Internet technologies. Furthermore, as worldwide communication links and traditional Internet access grows, markets will explode.

Figure 1.1 The Growth of Technology

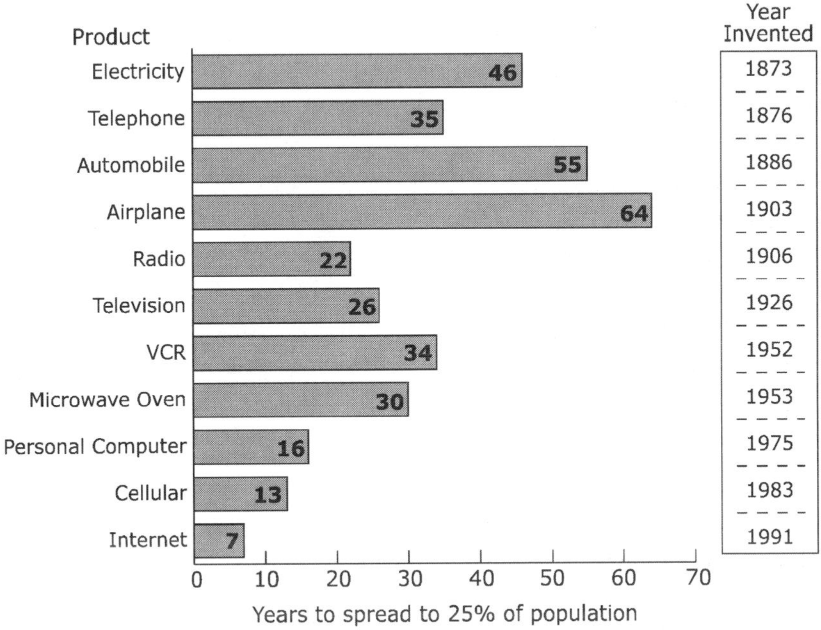

Product	Years to spread to 25% of population	Year Invented
Electricity	46	1873
Telephone	35	1876
Automobile	55	1886
Airplane	64	1903
Radio	22	1906
Television	26	1926
VCR	34	1952
Microwave Oven	30	1953
Personal Computer	16	1975
Cellular	13	1983
Internet	7	1991

Winners Think Outside of the Box

Selling on the Internet is not just creating a Web site and thinking "if I build it, they will come." Companies jumping on the next wave of marketing and sales via e-commerce think outside the box. Throw out the old business model and invent a new one.

Those companies that have been most successful and realized the highest return on investment have approached e-commerce using specific strategies. They rethink revenue streams, reengineer the business, empower customers, provide exceptional customer service, and join the global economy. Let's take a look at these five strategies to selling online, and how your company can take advantage of new methods of doing business.

Rethinking Revenue Streams

Companies approach e-commerce from every angle. A variety of methods of selling are being used in the Internet arena. For example, companies generate revenue from direct sales, online advertising, subscriptions, and credit card processing. Companies receive percentages of every online Internet transaction. Commissions are earned for matching buyers to sellers. Goods and services are auctioned online. Companies that generate revenue from direct sales include Charles Schwab and Dell. Companies that generate revenue from online advertising include Onsale.com and Yahoo!. The Wall Street Journal and TheStreet.com use a subscription-based model. VeriFone and CyberCash (credit card verification vendors) use transaction fees as a foundation for their businesses. Digital River and Beyond.com use electronic product delivery. Commissions for matching buyers to sellers compensate Realtors, as well as vendors whose sites feature clickthroughs to online stores, including Amazon.com.

To start on the journey, think about new streams of revenue. As you work through the book, you will see a whole new way of doing business and increasing your markets.

Reengineering the Business

The Internet promises to change your whole manufacturing process, allowing you to communicate instantly with suppliers, partners, and customers on a worldwide scale. The Internet requires an even bigger effort in business transformation, business reengineering. It is not enough simply to set up Web sites for customers and partners. To take full advantage of the Internet, companies reinvent the way they do business, changing how they distribute goods, and how they collaborate within the company and with suppliers.

Reengineering projects may be complicated, especially in the area of e-commerce. Since e-commerce is not merely a technology-based effort, it requires redefining marketing, sales, services, and products. In addition, it requires interactions between suppliers, shipping companies, legal departments, governments, and banks. It requires integration with existing systems either within the organization or between two companies. Successful companies have included marketing, sales, ordering, distribution, procurement, customer support, and trading partners in reengineering efforts. In addition, they have developed solutions that provide complete and timely reference data to company decision makers for marketing, product planning, production, and the logistics planning process.

Empowering Customers

The Internet provides a venue for buyers to have the most choices and best prices for products ranging from books to artwork. Even in business-to-business commerce, vendor malls allow buyers to purchase goods at the best prices.

Traditionally customers were limited by time, information and ability (distance) to find goods at the best prices and quality. Today, however, with a click of a mouse, customers are able to gather information about features and pricing, and perform comparison shopping with little effort. The result is the empowerment of the consumer, the driving force in the sales process.

Consider the traditional automobile purchase, which involves driving lot-to-lot, pushy salesmen, right price, wrong color, right color, wrong price. What if you could shop from home? One company that empowers its customers is Automall.com. This company functions as a virtual "one-stop shop" for its automotive enthusiasts by providing links to consumer information, manufacturers' sites, parts suppliers, financing and insurance services, and automotive accessories. Nearly 5% of all cars are sold on the Internet through sites such as www.automall.com.

Providing Customer Service

Customer service plays a critical role in e-commerce. Since e-commerce is not a market that geographically captures a customer, merchants must be even more creative in providing value-added services, such as online support, FAQ listings, follow-up e-mail messages, etc. The goal is to automate as many customer service processes as possible by leveraging Web technologies. If this is done correctly, your company not only generates cost savings, but also customer loyalty.

Joining the Global Economy

Your site should address the global economy. The Web is not just for the technology savvy. Women, men, and children of different countries, religions, and nationalities are participating in this new market. The bottom line is to make your site represent these areas by considering foreign languages, ease of use, and targeted interfaces. The advent of the global economy provides even more opportunities for those of you thinking about e-commerce to come and join this next wave of success.

E-Commerce Business Models

Industry divides electronic commerce into two main categories: business-to-business and business-to-consumer. It is important to understand these models, since e-commerce is approached differently, depending on if you are communicating with a business or with an individual on the Internet. In this section, we review both of these business models and provide examples of how companies successfully implement e-commerce solutions in both business-to-business and business-to-consumer commerce.

Business-to-Business

Business-to-business implies the selling of products and services between corporations and the automation of systems via integration. This category of commerce typically involves suppliers, distributors, manufacturers, stores, etc. Most of the transactions occur directly between two systems. For example, suppose that an aircraft company wants to build a plane. The plane requires parts from both large and small suppliers. The goal of e-commerce is to automate the entire supply chain. In this example, we call this automation "supply chain management" (the process of tying together multiple suppliers of goods to create the final product). The top half of Figure 1.2 illustrates a typical business-to-business e-commerce model. The model relates indirect suppliers, direct suppliers, transportation of the supplies, and their entry into the distribution system.

Figure 1.2 Business-to-Business E-Commerce

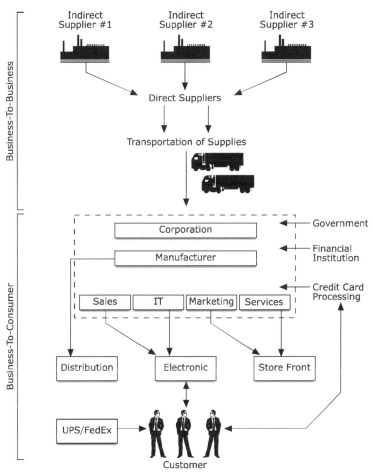

One company addressing the business demands of its suppliers is Daimler-Chrysler Corporation, a major U.S. manufacturer of vans, sedans, Jeeps, and trucks. Chrysler's suppliers of parts, packaging, and technology, now numbering over 20,000, rely on this manufacturer to communicate standards and share critical software applications. With the help of IBM, Chrysler developed and implemented the Chrysler Corporation Supply Partner Information Network (SPIN), an intranet-based supply chain management and support environment for distributing files over the Internet. Over 3,500 supplier locations access the Chrysler SPIN Web site. More than 12,000 users access information such as Chrysler's EDI Guide or QS9000 certification policies and procedures. They can also access dynamic database applications, such as real-time data, procurement analyses, and strategy applications. Productivity, driven by this solution, increased 20% within the first year of operation.

The Internet offers an excellent medium for transferring applications, time-sensitive information, and a variety of large files across a variety of platforms. It allows you to extend your enterprise, communicate with suppliers and buyers, and develop solutions that deliver as much information online as possible.

Business-to-business commerce also takes the form of malls. One example of a virtual mall, QCS, is an electronic community that helps retail buyers collaborate with their worldwide supply chain. While buyers manage their own back-end systems containing product information and transaction data, QCS offers a common interface for buyers to access different suppliers. This form of commerce parallels the retail mall concept and the architecture featured in some business-to-consumer enterprises.

Business-to-business commerce also takes the form of catalogs, another medium developed within business-to-consumer commerce. These electronic catalogs, used for purchases between companies, allow corporate buyers to search for products based on features or price. They present a single interface for individual sellers or multiple sellers of similar products. The primary benefits of these catalogs is their ease of use, flexibility, and easy updating.

The benefits to companies that succeed in business-to-business e-commerce are compelling. Business-to-business e-commerce, if done right, helps your company realize substantial cost savings, increase revenue, provide faster delivery, reduce administration costs, and improve customer service. These benefits result in a substantial return on investment. The most significant benefits are reduced administration, increased time for business, more accurate information, improved response time, and reduced errors.

Less time is spent on pushing paperwork, phone calls, faxes, and tracking all of this information. For example, the typical purchase order costs between $75 and $125 to process manually. With the improved automation that e-commerce offers, that cost can be reduced to about $3.

You can focus on increasing revenue through extending geography, improving sales channels, adding services, and growing market share.

Determining your current stock, shipment status of goods, and total costs becomes timely. Since information is the key to any business success, you can manage inventory on hand, shipment cost and methods, buying patterns, and distribution channels.

With automation, tasks can be performed instantly. If you consider the process of ordering an item, the order typically crosses a few desks, requires user input, and delays the transaction. With system-to-system communication, the amount of time it takes to generate an order can be reduced substantially. Automation reduces errors, because the process remains consistent across repeated transactions.

E-commerce between businesses is expected to be five times higher than business-to-consumer e-commerce. By 2003, Forrester Research Inc. estimates that business-to-business commerce will balloon to $1.3 trillion. Constituting 9% of all United States business trade, and more than the gross domestic product of either Britain or Italy, that's ten times the amount of business-to-consumer e-commerce. By 2006, that figure could represent 40% of all business conducted in the United States.

Business-to-Consumer

Business-to-consumer commerce involves interactions and transactions between a company and its consumers. Focus is placed on selling goods and services, and marketing to the consumer. Most people are familiar with the business-to-consumer e-commerce model. Companies such as Dell, Amazon.com, and eBay are becoming household names. Their main focus is to sell to consumers via the Internet. The bottom half of Figure 1.3 illustrates a typical business-to-consumer e-commerce model.

Figure 1.3 Business-to-Consumer E-Commerce

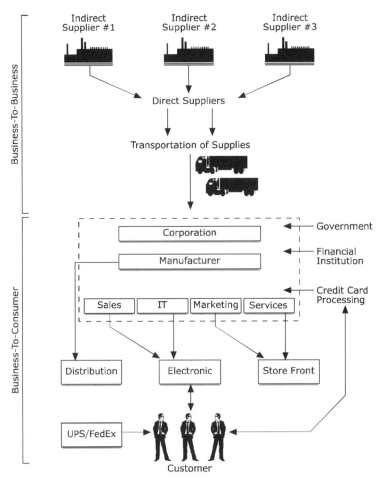

Think of the business-to-consumer e-commerce as the new QVC of selling your goods to consumers. Traditionally companies and small businesses employ catalogs, door-to-door sales people, telephone mail order, direct mailing and storefronts to sell products. E-commerce is the next medium to sell goods, using Web-based technologies. The key is to take advantage of these opportunities and do it right. And for the large corporations, it is time to rethink how you are currently doing business and how are you going to fit into the new global economy of e-commerce.

There are hundreds of e-commerce sites that are very successful in the business-to-consumer market. These companies search for innovative ways to sell products and services on the Internet. For example, a company called Infoseek allows its visitors to search phone books, yellow pages, and e-mail addresses online. Once you have found the person you are looking for, you can send a card, flowers, or call directly using a specific carrier. All of these transactions happen online.

Markets where convenience, pricing, and selection are major buying influences have done extremely well in business-to-consumer e-commerce. Computer software and hardware retailers have performed extremely well in the e-commerce market. Through electronic commerce, customers have a greater choice of products at lower prices than what they typically would experience while shopping at major retail outlets. Companies such as Compaq and Dell have added millions of dollars in sales per day. Travel agencies also represent key players in the electronic commerce industry. Consumers can now browse the Internet for the best fares online without having to call several agencies or wait on the phone for quotes from each airline. They can now, with a click of the mouse, search for pricing and dates of airfares, hotel rates, and vacation packages. Other businesses that have flourished include book and magazine retailers, music and video vendors, and retail florists.

Often we are asked, "Where is the personal touch in e-commerce when selling products to consumers?" In today's society, people are looking for convenience. Traditional selling will not be eliminated, just approached differently. Let's say for example that you buy the same t-shirt and socks every year. These items are standard and do not require that you go to the mall and select them. Going to the store may be more of an inconvenience than pleasure when selecting these commodity items. As a merchant, why not put these items online and allow your consumers to purchase these items from home? Customers save time and money. In addition, retailers save time and money. Electronic commerce provides a compelling avenue for selling products to consumers. The key is doing it right. Sell the right product, at the right price, to the right audiences.

Business-to-consumer e-commerce sales provide extraordinary opportunities for most corporations. Businesses have the opportunity to sell products and services 24 hours a day, to reduce costs associated with retail space, personnel and supplies, and to increase market share. Small businesses can play with the big boys. The most significant advantages include higher visibility, branding opportunities, direct revenue generation, attraction of new customers, and worldwide exposure of the business.

E-commerce provides an additional or even primary venue for companies to advertise goods and services. The limitations placed on traditional selling of goods by location have been virtually eliminated. The Internet is a worldwide marketplace and should be approached accordingly.

E-commerce allows you to create your own image or change the public perception of your company. What will distinguish your e-commerce storefront is your creativity in presenting a storefront that is appealing to your customers worldwide. Some of the biggest successes on the Internet are less than five years old and have branded themselves as household names. They have focused on marketing, customer service, site design, and selling products and services which keep their customers coming back for more.

E-commerce, with 24-hour sales and support, improves revenue by featuring a store that never closes. E-commerce truly incorporates the concept of making money while you sleep. You do not have to physically be available 24 hours a day even if your customers are buying 24 hours a day. Isn't that a refreshing change?

E-commerce reaches customers beyond your traditional marketplace. This is the opportune time to start expanding, not your typically brick-and-mortar storefronts, but your presence on the Internet. As more and more consumers get online, they will be looking for your products and services. Capture these new customers by deploying a strong and compelling e-commerce solution.

E-commerce enables 24-hour support. A Forrester Research, Inc. study of financial institutions estimates that Web service costs companies just four cents per customer on average for a simple Web page query, vs. $1.44 per phone call. Shifting service to the Internet, notes Forrester, could allow companies to handle up to one-third more service queries at only 43% of the cost. This not only saves you dollars, but provides additional convenience for your customers to get the help they need immediately.

E-commerce generates a global customer base. One location can distribute goods worldwide. Not just a market for your town, state, or country, your company enters the global economy. What a change!

How Businesses Change

Let us take a look at two business models that describe how companies are currently receiving goods from their suppliers and the channels that they use to deliver these goods to the consumer. These models provide a basis for understanding how your role, whether you are part of a large company or have a small business, may change as industry moves rapidly towards the e-commerce model. Each company will be dramatically affected by their revised business model.

Traditional Selling Chain

Traditionally, selling to the ultimate customer (the consumer) has required that suppliers and manufacturers send the goods to the distributors. The distributors in turn sold the goods to the wholesalers or resellers. Finally, these goods are sold to the ultimate consumer. Each person in the chain added their service fees to the products. In the end, the costs and time spent on the product increased dramatically as each entity in the chain handled and distributed these products. The diagram below (Figure 1.4) illustrates the traditional selling chain.

Figure 1.4 The Traditional Selling Chain

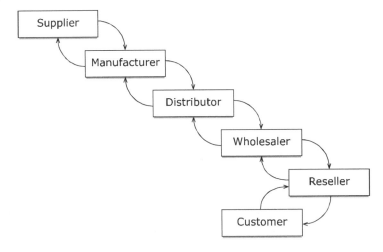

Direct Marketing Selling Chain

Direct marketing strategies break the traditional selling chain. Suppliers and manufacturers can bypass a variety of intermediates in order to sell directly to the customer. Note the difference between Figure 1.4 and Figure 1.5. Industries such as airlines (selling tickets without travel agents), online bookstores (selling books without traditional brick-and-mortar components), stock brokers (processing transactions online), and computer manufacturers (offering hardware and software via the Internet) have moved toward this model of selling goods on the Internet. Not only have they realized cost savings through this model, but have passed these savings to the consumer.

A word of caution, remember that intermediates do add value to products through their specialized services. These added values might be lost or reduced within a disrupted selling chain. Determining where value is added in the selling chain enables companies to maximize the value of the selling chain disruption.

Figure 1.5 The Direct Marketing Selling Chain

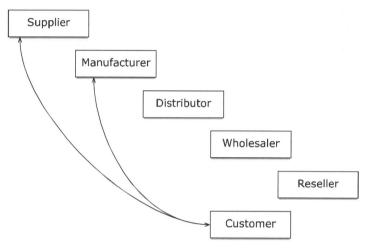

In reviewing these two models, determine how your business fits in this new market. Although the resellers and retailers are still selling goods online, this may change as suppliers and manufactures find venues to start selling to their ultimate customer, the consumer. As e-commerce changes traditional selling, think about how your business will change.

Gaining Competitive Advantage

How do I get started? This section provides a starting point for determining how your business fits into the e-commerce economy. Keep in mind when deploying an e-commerce solution, that it is important to understand your business before thinking about the technology required. E-commerce is not a technology-driven business. E-commerce simply amplifies business. E-commerce is one of the first areas in which a business can gain a return on investment by deploying a sound solution. And this return can be substantial.

In the deployment chapter, we cover the entire design, development, and deployment process from defining the vision statement to deploying the technical e-commerce solution. However, to get you started, this section provides the basis for developing an e-commerce strategy.

To get professionals started in thinking about ways to use e-commerce in their companies or for private business, we offer two exercises. Take out a piece of paper and a pencil. These two objects form the starting point for your e-commerce solution (whether for your company or your personal use). In addition, as you work through the book, keep a notepad nearby so you can continue to build your solution.

The first step to building an e-commerce solution is to determine your vision, goals, and strategies. Once you know where you are going, brainstorm on ways to get there. At this point, do not think about the technology. Instead, think about how you can take advantage of the Internet to sell your goods and services. It is now time to think outside of the box.

Vision Statement

The vision statement sets the tone for the entire project. It must be compelling. For example, one company wants to develop a one-stop shop for exotic hardwoods. Another company wants to harness their current strengths by selling training materials online. Unless your plan calls for you to compete only on price, try to find a niche that completely integrates your dreams with your offerings to the world.

If you struggle with the concept of a vision statement, perform the first exercise. Spend the next week or so carrying around a notepad. Every time you experience something inconvenient, whether you are at work, at home, or at play, write it down. Study the problem. Can it be addressed by harnessing the power of electronic commerce?

Goals and Strategies

A goal is a measurable account of "what" you want. A strategy is a technique for achieving your goal. Set your goals for the e-commerce solution and define specific strategies to attain those goals. Goals and strategies are required in order to help ensure that you live up to the vision statement. Furthermore, when you are defining your new methods of marketing and sales, you can map all decisions back to the specified goals and strategies.

Brainstorming

Now it is time to think outside the box. Develop a plan that moves you toward the company vision, goals and strategies. Use the second exercise as a basis for defining your e-commerce solution. You may also want to refer to the project deployment chapter to further define your plan as you read through each chapter in the book. The key to the exercise described in Table 1.1 is to consider your competitors, your suppliers and customers, and your current business process. How are your competitors, direct and indirect, doing business electronically? By researching the efforts of the competition, at least you establish the minimum expectations for the solution. List your suppliers and buyers. Determine how the business currently sells goods, provides services, and markets its wares, in either case.

Table 1.1 Current Business Process

Your Customers	Your Suppliers
How are we doing business with our customers?	How are our vendors doing business with our company?
How do we market to our customers?	How do our suppliers market to our company?
How do we sell to our customers (catalogs, online ordering systems, toll-free calls)?	How do our suppliers sell to our company (EDI, catalogs, online ordering systems, toll-free calls)?
What services do we offer our customers (24 hour phone support, return policies, etc.)?	What services do our suppliers offer our company (24 hour phone support, return policies, etc.)?

Redefining your Business Process

Consider new ways of conducting business in the e-commerce arena. How can we address these areas more creatively, more efficiently, and more consistently? Answer the questions posed in Table 1.2 with a focus on your revised business process.

Table 1.2 Revised Business Process

Your Customers	Your Suppliers
How should we do business with our customers?	How should our vendors do business with our company?
How should we market to our customers?	How should our suppliers market to our company?
How should we sell to our customers (catalogs, online ordering systems, toll-free calls)?	How should our suppliers sell to our company (EDI, catalogs, online ordering systems, toll-free calls)?
What services should we offer our customers (24 hour phone support, return policies, etc.)?	What services should our suppliers offer our company (24 hour phone support, return policies, etc.)?

After you have redefined your business process, match the new process to your vision and goals. It is critical that your new business model is in line with your corporate goals and strategies. In order to maintain the integrity of your solution, do not expend energy in areas that do not add value. Often companies head in a direction that does not add directly to the their bottom line. By mapping your new business process to the corporate vision, goals, and strategies, you ensure that each new process brings immediate value to the company.

2

Sales and Marketing

- Assuring that your e-commerce site is a success
- Marketing strategies for e-commerce
- Attracting and keeping customers
- Sales strategies for e-commerce

Marketing your Web site is critical to the success of your e-commerce endeavor. Marketing leads to visits. Visits lead to sales. Sales lead to repeat sales. If you like the result of this process, repeat sales, focus on the raw material, your marketing effort. The most successful e-commerce companies place a major emphasis on marketing their sites. These sites take advantage of both traditional media, such as magazines, newspapers and TV commercials, and electronic media, such as banner ads, search engines, and e-mail distributions lists. The same level of effort you devote to your traditional advertising must be put forth when marketing your business electronically.

Whether you currently have a site (or plan on implementing a site in the near future), this chapter surveys the main issues that you should consider in order to exploit various media opportunities. Depending on your budget, some forms of advertising may be too expensive to deploy immediately. On the other hand, some may prove less expensive (even free) than your current methods of selling your products and services in both business-to-consumer and business-to-business e-commerce.

As you explore this chapter begin thinking about how to immediately apply these sales and marketing techniques to your current business. Ask yourself these questions:

- How am I marketing my products and services?
- How can I add Internet-based marketing to my current marketing strategies?
- What are innovative ways to market on the Internet?
- What industries or companies would prove beneficial for me to list my products or services on their sites?
- What industries or companies would benefit by listing their products or services on my site?

Use these questions as fuel for a brainstorming session. Think of all of the options, even if they seem out of reach. Remember, companies that are successful in e-commerce are thinking outside the box. Once you have considered all of your options, prioritize the items based on which medium delivers the most value in the shortest amount of time (the most bang for your buck). Your goal is to attract customers to your site and initiate the sales process.

After listing all of the ways you want to market your products and services, develop a plan of action. This plan details how you intend to target each major marketing area, including a specific time frame. If you or your company wants to work with specialists in the area of e-commerce sales and marketing, we have provided a list of major companies that offer sales and marketing services at the end of the chapter. However, even if you use a third-party marketing company, know what to look for and what questions to ask. If you learn the components that comprise a marketing plan, you will have a better understanding of your options.

The Internet itself also offers some excellent sites for finding additional information about Internet marketing. Some of these marketing sites include www.adresource.com, www.internet.com, and www.net2b.com.

More Marketing? More Marketing!

A considerable number of companies, even the smaller ones, have realized a substantial growth in sales. How are they doing it? Companies are spending a considerable amount of resources on online marketing. A report by Simba Information estimates that online ad spending will reach $7.1 billion by 2002. The report also states that the challenge for Web publishers is to hold the audience captive long enough for them to absorb the advertiser's message. Companies use a combination of traditional marketing methods, such as magazine ads and press releases, as well as online advertising at major portals such as Yahoo!, Lycos, and Excite.

For example, sales for Camera World with a site at www.cameraworld.com grew from $1 million in 1997 to over $16 million in 1998. The company sells analog and digital cameras, high-end and standard camcorders, digital camcorders, VCRs, and DVD players. For advertising, they experimented with some banner advertising and search engine listings. But what seemed to work best for them were print ads in Popular Photography, Petersen's Photographic, Shutter Bug, PC Photo and various other camera or electronics-related magazines. Camera World is also forging partnerships with other online retailers.

K&L Wine Merchants at www.klwines.com sells wines and liquor. Their most effective advertising is word of mouth. The company also placed ads on Yahoo!, Lycos, and Sidewalk. In September of 1997, they were named by Money Magazine as the "Internet's Top Wine Site." The resulting publicity led to a boom in business.

Larger companies such as Amazon.com Inc., E*Trade Group, and eBay have focused on becoming brand names via the Internet. They have placed emphasis on marketing their site through trade magazines, press releases, banner ads, TV commercials and other forms of marketing. The result is that they flourish among the top e-commerce sites on the Internet. Amazingly, most of these larger e-commerce sites were unknown just three years ago.

Strategies to Attract Customers

Now the question is "where to start?" Sales and marketing strategies on the Internet involve three basic questions:

- How do I attract customers to my site?
- How do I keep customers at my site?
- How do I track customer purchasing trends?

This chapter answers all three questions. The strategies discussed in the following sections apply to any company with a presence on the Internet. As you learn about each strategy, consider how you might apply it to your own business. The strategies include:

- Search Engines and Directories
- Advertising Banners
- Online Classified Advertisements
- Message Boards
- Registering Users and Sending E-mail
- Links on Other Web Sites
- Newsgroups
- Discussion Lists
- Traditional Media
- Press Releases
- Trademarks and Branding

How Will They Find You?

Georgia Tech University recently conducted a survey in order to determine how a variety of people discovered specific Web sites. In order to gauge the popularity of various media, both traditional and online, the survey allowed participants to respond with multiple answers. The results show that a majority of the respondents are directed to specific Web sites by search engine listings, by links from other sites, or by advertisements in printed media. Refer to Figure 2.1 for details of the survey.

Figure 2.1 How People Discover Web Sites

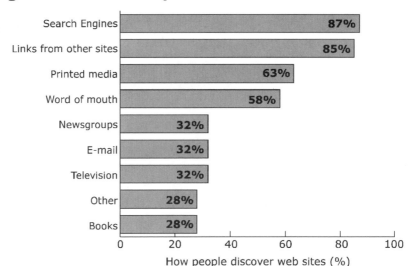

How people discover web sites (%)

Search Engines and Directories

Search engines and online directories, both extremely cost-effective, prove most popular as marketing opportunities. They catalog and list your Web site information so that when someone using the Internet searches for information pertaining to products or services that you sell, your potential customer locates your site. Search engines and directories function as listings of your site's theme and content, similar in function to a phone directory advertisement.

As soon as your Web site is up and running, you should list its contents with several of the major search engines and directories or with those that are specific to your industry. Hundreds of lists are available to display the particulars of your company. Some of the major search engines include Yahoo!, Lycos, and Excite. Industry-specific sites that list your business, although typically among your competitors, provide you with an opportunity to make your information public. Most of these sites provide free listings of your site. The information you provide for these listings varies, whether for a major commercial or an industry-specific site.

Although the distinction between the two is often con-
fused or ignored, search engines and directories are the
products of two entirely different processes. The processes,
described below, also affect your Web site listing when you
alter your site. If you, the Web site designer, edit your Web
pages, the search engine technology reflects the changes following the next search request. However, for directories,
editing your Web pages does not update your site listing.

Search engines create their listings with minimal human
intervention. They crawl across (search) the Web to gather
information from existing Web pages. Page titles, body copy,
and other elements each play a role in the resulting collection of data. The captured results display in a logical structure so the initiator of the search (the potential visitor) can
view each item on the list. A popular search engine that
uses this technology is HotBot (www.hotbot.com).

Directories require that the Web designer (or other company representative) submit a form containing keywords
that describe your site. When a user enters a keyword for a
search, a directory lists those sites whose authors have submitted that keyword as part of the search criteria. A popular
site, Yahoo! (www.yahoo.com) uses directories as a method
for categorizing information on their site. Look for a description of the registration (or listing) process for Yahoo! later in
this chapter.

Whenever you use a search engine to search the Internet, notice that you receive a list of sites ranked according to
relevance, that is, how closely the site matches your entries.
How are these sites ranked from the first item to last?
Search engines follow rules that describe relevance according to the location of words and their frequency. Search
engines check to see if the keywords appear near the top of a
Web page, such as in the headline, subtitle, or in the first
few paragraphs of text. Web pages with keywords appearing
in the title are considered more relevant than others to the
selected topic. In addition, search engines analyze how often
keywords appear in relation to other words in a Web page.
Those with a higher frequency are often deemed more relevant than other Web pages.

Every search engine available on the Internet supports both the location and frequency method for listing and ranking sites. However, some sites have added other criteria in order to rank pages. The differences in the criteria explain why the search for an item on one search engine results in different matches than on another search engine.

Examples of this situation abound. The search engine, Excite, uses link popularity as part of its ranking criteria. It determines which Web site pages in its index have a multitude of links pointing at them. HotBot and Infoseek give a slightly higher ranking to pages with keywords in their metatags. Some search engines reduce the ranking of pages or do not list them at all, if they detect search engine spamming. This type of spamming involves repeating a word multiple times in an attempt to generate a higher ranking for a page. Do not expend too much energy learning the distinctions between search engine ranking methods. Just remember that there are legitimate reasons why searches on different search engines generate different results.

Begin your marketing efforts by registering your site with some of the major search engines and directories listed in Table 2.1. You may also want to register your site by employing venues specific to the products and services you are selling. Note that Yahoo! and several other engines generate searches specific to certain regions, states, cities, and countries.

Now you have a relatively comprehensive list of the major search engines and directories. The most efficient means of exploiting these sites is to list with sites that generate the most traffic. Research indicates that almost half (46%) of all Internet traffic generated by search engines and directories is a byproduct of Yahoo!. Excite accounts for about 22% of the traffic generation. AltaVista accounts for almost 10% of the traffic. These three venues comprise three-quarters of the Web's traffic generated by search engines and directories. Begin your marketing work with these three sites, but do not ignore the specialty sites.

Table 2.1 Search Engines and Directories

Name	Address	Profile
AltaVista	www.altavista.com	One of the largest search engines on the Web, a favorite among researchers
Direct Hit	www.directhit.com	Ranks sites based on the number of hits received
Excite	www.excite.com	Offers a medium-sized index and integrates non-Web material, such as company information and sports scores
GoTo	www.goto.com	Companies pay to be placed higher in GoTo's listings
HotBot	www.hotbot.com	Favorite among researchers due to its large index and many search features
Infoseek	www.infoseek.com	Offers small to medium-sized index, but includes person listings and company pages
Lycos	www.lycos.com	A directory model
MSN	www.msn.com	Features directory listings and search engine results
Netscape	www.netscape.com	Offers a branded version of the Excite search engine
Search.com	www.search.com	A branded version of the Infoseek search engine offering specialty searches
Web-Crawler	www.Web-crawler.com	Offers the smallest index of any major search engine
Yahoo!	www.yahoo.com	Largest human-compiled guide to the Web with over one million sites listed

You can register your own site or you can employ a commercial service to do the work for you. In order to register your site, review the seven marketing sites that follow. DirectoryGuide (www.directoryguide.com) features over 350 search engines and directories, each organized into distinct categories. MMG's WebStep Top 100 (www.mmgco.com/top100.html) allows you to list your Web site on the top 100 search engines. The Submit Site (www.thesubmitsite.com) and WebCom (www.Webcom.com/html/publicize.html) also allow you to register your site. Submit-It (www.submit-it.com) allows you to post your Web site details to more than 20 listings from a central location. SelfPromotion.com (www.selfpromotion.com) also submits sites free of charge. If you need further information on search engines, WDVL's How in the Web will they Find Me? (www.stars.com/Search/Promotion) offers instruction for designing a site, generating publicity, and using search engines.

In order to use a commercial service to list your site, take a look at Yahoo! and Express Press. Yahoo!'s announcement services provide links to guides and to organizations that you can hire to help publicize your site. User information is available within Yahoo's Web site. An alternative to Yahoo!, Express Press (www.xpresspress.com) distributes e-mail news releases to hundreds of editors throughout the world, targeted by subject area.

To gain a solid understanding of the registration process, walk through a typical registration at Yahoo!.

1. Determine whether your site is already registered. Check by searching Yahoo! for your site by title or by browsing the appropriate categories. This determination is important because if your site is already listed, your mission is complete.

2. If your site is already listed in Yahoo! but you want to change its comments, title, and/or placement, use the change form.

3. Specify the appropriate category for the listing of your Web site.

4. Click the hyperlink to display a list of important pointers explaining how the directory is organized. Finding an appropriate category for your site is at the heart of this process. Remember that Yahoo! Surfers visit each site suggested, and proper categorization on your part helps Yahoo! process suggestions quickly.

5. If you are suggesting a scheduled chat, live broadcast, or chat area, please submit directly to Yahoo! Net Events.

6. If your site is in a language other than English, please check the list of non-English directories to see if there is an appropriate International Yahoo! for your submission. Sites that are in a foreign language represented by international directories are not added to www.yahoo.com.

7. Suggest your site from the category you think most appropriate by clicking on the "Suggest a Site" link at the bottom of that category page. Remember that the final placement of your site is determined by Yahoo! Surfers. Suggesting a site is free.

8. Click on "Suggest a Site" in the appropriate category. An online form displays, requiring information about your Web site.

9. Provide, among other information, your site's title, URL, and a brief description.

A majority of the search engines and directories are easy, inexpensive, and effective to use. However, ensure that you include all the major keywords for directory listing sites and place your keywords and titles on the page so that search engines will rank them higher on their lists.

Advertising Banners

You jump to a new Web site. Not much of interest on this page. As you prepare to jump somewhere else, you notice a small (two inch by one inch) advertising banner screaming "Click Me." You click the banner...and jump to an entirely new site. The banner has done its job. Banner ads display on a Web page when a potential customer accesses the site. Clicking the banner links visitors to your e-commerce site.

Banner ads are one of the most blatant methods of marketing goods and services over the Internet. Research suggests that online banner advertisements perform as effectively as television advertisements in their ability to increase consumer awareness of brands. In addition, the recall level of a banner ad is equal to that of a television ad. So even if the banner ad can not coax you to click through it, you still may take its message with you. One reason that banner ads are so successful is that Web site visitors seem to understand that banner advertising may reduce the amount of money they have to spend for content.

Banner ads can be placed on any site that you think will attract the most customers. For example, if you sell car insurance, you may want to place an ad banner on www.automall.com. When customers buy a car they also immediately purchase automobile insurance. Or you may decide to place a banner at a bank's Web site, so that when people apply for car loans they can apply for insurance. To determine where to employ your banner advertisements, consider site content, placement, and cost.

Who is your customer and what do they typically buy? Suppose you sell fishing gear. Your goal is to place banner ads on sites that attract fisherman. Answer the following questions. What are the demographics of my audience? For fishing gear, the typical buyer is male, ranging in age from 15 to 65. What Internet sites cater to this market? Would a visitor who searched on the keyword "recreation" be interested in my products? What traditional media can I employ to display my banner? Would a banner help my business if I placed it on the Web site of a national or regional magazine?

A beneficial exercise involves brainstorming to generate all of your options. Then choose those sites with the greatest number of potential clickers (visitors). One technique for creating a list of candidate sites involves conducting a search for your product or service using a search engine. Put yourself in your customer's shoes. Think of how they might search the Internet. For example, go to www.yahoo.com and type in the name of your product, such as "fishing gear."

Review the resulting list of sites. Visit the ones that seem relevant and determine whether they offer banner placement. Typically each site lists a contact number for banner placement opportunities. Some sites allow you to submit banner ads directly from a Web page.

Once you have decided which sites to use for banner ads, you can decide where to actually place the ads. Although they can be placed anywhere on a page, ads prove more successful when they are appear near the top of the page. The page top ad delivers higher clickthrough than the same ad placed at the bottom of a page.

Once you have considered your options for placing banner ads, take a look at your advertising budget. Some of the major sites charge a substantial fee to place your banner ad on their site. Others may charge only a small fee.

Web advertising is typically sold on a cost per thousand impressions (CPM) basis. An impression occurs when a visitor to a Web site views a page where an ad is displayed. Banner ads are typically sold in quantities of 100,000 impressions. The cost may vary from $20 to $100 per thousand impressions (popular sites average $25-$70 CPM). Because most sites have repeat visitors, your 100,000 impressions may be generated by only 20,000 to 50,000 unique visitors. Typically, software packages allow you to specify not to display a banner more than once during a single user session.

Most popular sites offer ad agency discounts and volume or frequency discounts (typically 15% below the gross rate). Some Web sites also offer cost per click or clickthrough. This type of advertising provides a direct measure of response and ad performance, as you are billed according to the number of times a user clicks on the banner ad. In addition, this information can be easily tracked for analysis purposes. You can determine which sites performed best, and for which categories, search words, and Web pages. Although these advertising arrangements are available, most sites avoid selling this way. Poorly designed ads do not attract clickers, so ad revenue is reduced.

Some Web sites generate revenue from advertising fees. Sites such as Onsale.com rely on banner ads and other ads for a majority of their revenue. You may want to consider using your site as a host for other company's advertisements. However, in some cases, banner ads for other sites prove problematic. For instance, you may notice that most major department stores, hardware vendors, and other large companies do not place advertising from other sites on their site. These companies want customers to focus on buying their products, instead of being distracted by other company banner ads.

Depending on your budget, banner ads can be a great way to market your products or services. When assessing their value, consider the sites that act as hosts, the placement of the banner ads on a site, and the expense associated with each.

Online Classified Advertisements

Several sites allow you to place classified ads for free or for a minimal fee. For example, Yahoo! has a classified section to place items you want to sell for free. Other major search engines also provide this service. Classified ads can be placed in several sites. For example, there are thousands of real estate sites that allow you to place ads to buy or sell homes or investment properties for free or at a minimal cost. These sites make their revenue from advertising. Their goal is to get as many customers as possible so they can charge other companies such as real estate, insurance, and mortgage companies higher fees for advertising on their sites. Also, most local newspapers and magazines are online. You can place your ads on these sites at a minimal cost. For example, the Baltimore Sun allows customers to place ads at $1 per week. This is much less expensive than newspaper ads that can cost up to $80 per ad.

Message Boards

Special interest boards often serve as listing areas for product or service providers. You can list your particular offerings on several message boards across the Internet. These message boards tend to address special interest groups, such as photographers, pet lovers, cooks, etc. To locate these boards, search on any of the main search engines for keywords related to your products or services. In addition to message boards, you can supplement your listings by using special interest malls. These malls result from a collaborative effort among multiple vendors targeting the same special interest groups.

Registering Users and Sending E-mail Messages

If your site has been well-designed, you should have a list of visitors and customers who provided their e-mail addresses to request notification of future developments and special offers. This list of names is a valuable resource and should be treated accordingly. You can send e-mail messages to announce sales on products, new products, or information about select products.

You can design the forms on your Web site, by including major categories of interest, so that site visitors or customers can select the information they want to receive. For example, if you are selling clothing, you could set up your site so that visitors can choose to receive information about specific designers. If the designer announces a new line of products or if you have a sale on a designer's line, you can send your potential (or repeat) customer an electronic mail (e-mail) message.

User registration and e-mail technology should combine for proactive selling. For example if you are selling gifts, you could set up an online form to allow customers to enter information such as birthdays of their family members or friends. After processing this information, you could send your customers reminders to purchase gifts on those dates. By capturing the date, you generate more sales. By reminding your customers about the dates, you provide a convenient value-added service.

E-mail messages are a great medium for creating repeat business. Consider an online pool supply store. One of this company's best selling products is chlorine for swimming pools. A customer replenishes his chlorine supplies on a fairly regular schedule. The Web site generates an e-mail message to the customer requesting a chlorine supply reorder automatically. The customer can respond by clicking an element, usually a button, imbedded in the e-mail message to place the order. Without this functionality, this same customer would have to remember that he needs to reorder chlorine, locate the site on the Web, locate the order section, and then input his order information. The improved process ensures convenience for both the consumer and the business...a definite win-win situation. What implementations can you use to ensure win-win situations?

When e-mail is used as a medium for advertising, targeted e-mail messages are much more effective than scatter-shooting. If your customer receives e-mail from your company that is of interest or somehow adds value, you have created a positive experience. On the other hand, if your messages are too frequent or impart little value, you create a negative experience. At best, your message is deleted. At worst, your potential customer feels irritated, and denigrates your products or services. When advertising via e-mail messages, consider the guidelines that follow.

Do not reward your customers for registering at your site by spamming them. Spamming means sending them incessant e-mail messages whether requested or not. To avoid sending e-mail to people who are not interested, ask them on your Web site if they want to receive e-mail from you.

Treat the information a visitor leaves at your Web site as confidential. Typically, sites inform users that they collect data for internal purposes only, such as analyzing sales trends. Respect the privacy of your visitors. They may become your customers.

If you plan on selling the information that you gather, secure the visitor's permission. Have you ever received faxes in the middle of the night or useless mail from marketing companies? Companies also employ e-mail messages as a means of selling products and services. Mailing lists are being purchased and sold across the Internet. If you sell visitor information, and it results in annoying or incessant e-mail to your customers, you have violated their trust. Do not trade long-term trust for short-term revenue.

Follow up the sale of a product or service with an e-mail message. Often the only time that you become aware that a customer had a bad experience is after that customer vanishes. Angry customers do not complain, they disappear. Give all of your customers the opportunity to rate their buying experience. By following up with an e-mail, you can capture potential problems of customers. You can include a questionnaire with a few questions asking about their shopping experience. Or you can follow with a telephone call. Use any resulting information to improve your site. Remember that customer service is even more critical on the Internet, because your competitor is just a "click away."

Links On Other Web Sites

One of the first things we needed to do when advertising this book was to determine all of the sites where we would place links to our site. We approached this task by putting ourselves in our customers' shoes. This means thinking of all the sites that interest your audience. Where do your customers go when they surf the Web?

When you have identified a variety of sites, negotiate a cross-link with their owners. It may be much more difficult negotiating a cross-link with a site that has a greater traffic level than yours. If you have a great negotiator on staff, make an effort to persuade the owners. Often you can sell the unique benefits of cross-linking with your site. It never hurts to try. There are lots of alternative sites to target.

You might also decide to compensate certain Web site owners for placing your link on their site. Some companies offer a percentage of the sale if a customer purchases a product via the link. For example, Amazon.com offers Web sites 15% of a book sale when a customer purchases a book via a link to Amazon.com. The Yahoo! Store includes built-in tools to help you create and manage revenue-sharing links. Use the same strategies as those for banner placement, discussed earlier in the chapter.

Newsgroups

About five years ago, when the Internet first became popular, two attorneys advertised their services by posting to every existing newsgroup. There are thousands of special interest newsgroups on the Internet. The furor from their postings resulted in angry newsgroup users sending several thousand e-mail messages to these attorneys in complaint. The attorneys had violated the non-commercial nature of these newsgroups. As a result of their scattershooting, the Internet service provider (ISP) unceremoniously dropped the attorneys as customers.

Newsgroup topics range from technology forums to gourmet food to aliens attacking the planet. Posting to these newsgroups is free. Determine which newsgroups might have participants interested in your products. Post information that is informative to a group. In order to use newsgroups as an advertising medium, include a signature file listing your uniform resource locator (URL) address. Do not directly advertise on newsgroups, but concentrate on finding ways to add value to the discussion and create interest among the participants.

Discussion Lists

You can subscribe and post to discussion lists in order to exchange information about various subjects. Instead of sending e-mail throughout the day, hosts of these lists compile the e-mails in digest form and e-mail them infrequently. These discussion lists represent a compelling medium for marketing to potential customers, but we advise discretion. Create value for your fellow participants. Ensure that your exchanges are win-win.

Table 2.2 lists a variety of resources for using discussion lists effectively. Review each of these resources before developing a strategy to target discussion lists and their audience members.

Table 2.2 Discussion List Resources

Name	Address	Profile
Internet Advertising Discussion List	www.internetadvertising.org	A non-profit, sponsor supported discussion group
E-mail Discussion Group Resources	www.Webcom.com/impulse/list.html	A directory of discussion lists and resources
Internet Marketing List Archives	www.i-m.com	No longer active, but archives are available
L-Soft Catalist	www.lsoft.com/catalist.html	The official catalog of discussion lists
List Exchange	www.listex.com	A list of discussion lists and resources
Guidelines for Usenet Newsgroup Creation	www.Web.presby.edu/~jtbell/usenet/newgroup/guidelines.faq	A set of guidelines for creating newgroups in the Usenet newsgroup hierarchy

Traditional Media

Both small and large companies use traditional forms of marketing to publicize their businesses and to attract people to their site. Onsale.com launched a major television advertising campaign in order to publicize their Web site. Automall.com also uses television advertising to entice visitors to their Web site. As a starting point for your advertising efforts, place your URL on the following media: business cards, customer mailings, letter heads, e-mail messages, Yellow Page ads, and any other medium that you currently use to market your company. You may also want to advertise your products and services in the newspaper and their classified ads, in trade magazines, in television commercials, and in infomercials.

Most people still have not purchased online and may not be aware of what products or services are available. As more and more people make purchases via the Internet, online advertising becomes a more viable method of advertising. Your goal is make people comfortable purchasing your products online. The first time a transaction is completed successfully, the next one is easier. You just need to get your customers started.

For example, we suggested that a major department store should consider auctioning their products online. We recommended that they place an ad in the newspaper, just as they would for a big sale. This advertisement would attract potential customers to the site because of the curiosity created and for the potential deals available. Once these customers became familiar with the process of the online auction and the resulting auction prices, they were hooked.

As you survey traditional media, consider how businesses support their online efforts and how you can incorporate your online message. In a recent study conducted by the Response Marketing group, Web addresses, used as direct respose mechanisms, appeared as frequently as toll-free phone numbers. They were both used by 58% of the sample in the study. Street addresses (17%), "bingo" cards (4%), e-mail (4%), and coupons (1%) were also used as direct response mechanisms in the sample. The industries most likely to use a URL in their ads were real estate, water, and delivery services, followed closely by computer businesses, telecom organizations, and office equipment companies. The study also found that URLs are most often used as response mechanisms in computer magazines (93% of ads), news magazines (88%), and financial magazines (79%). The additional cost of adding a URL to your traditional advertising is minimal, but invaluable.

Press Releases

What is unique about your Web site? Develop a compelling reason to visit your site. Send a press release to your trade magazines, local newspapers, online magazines, or business magazines. In addition, some Internet sites allow you to submit stories that might be of interest to their readers. If you are not up to the task, hire a marketing company to help. We have often contacted local newspapers and magazines to promote our services by discussing topics of interest, including e-commerce or other computer technologies. If a newspaper or magazine editor finds your site compelling, so will the readers. Again focus on adding value and on generating interest.

Trademarks and Branding

Is Kleenex the only form of tissue? Is Band-Aid the only form of adhesive strip? Is Coke the only form of cola? Branding is a powerful tool. Trademarking is something you do to the public. Branding is something the public does to you. When we started our company, the first thing we did was heavily trademark our product, capturing the ideas that allow us to add value. We put a trademark on the concept of "Architecting the Solution" and the "E-commerce Program" (discussed later in the book). We also associated the red "E" with our name and services by trademarking it. Your goal is for people to think of your brand when they are purchasing a particular product.

A recent industry study says the top 100 e-commerce sites spent an average $8.6 million each in 1998 to build their online brands and attract customers to their Web sites. Even the smallest companies in this group, those with 100 or fewer employees, spent an average $2.2 million on marketing. The largest companies in the group spent $21.4 million each (on average). The top five marketing budgets in 1998, according to the report, belonged to Amazon.com ($133.0 million), E*Trade Group ($71.3 million), Barnesand-Noble.com ($70.4 million), CDnow ($44.6 million), and Ameritrade Holding Corporation ($43.6 million). Wow!

You may ask why they are spending so much. The reason is they want to be the front runners (the brand names). Books online? Amazon.com. Trading online? E*Trade. Compact disks online? CDnow. The principals know that if they brand early in the e-commerce wave, they can capture future markets. Remember the statistics, revenue will reach

$2.2 trillion or more in the next four years. If you can capture and maintain even a small percent, imagine.

Strategies to Keep Customers

So you attracted customers to your site with a combination of marketing and advertising techniques introduced in the previous section. Now your questions are "how do I keep them here" and "how do I get them to buy my products and services?" This section describes five fundamental strategies to generate customer loyalty and to create buying incentives during the first visit, the second visit, the third visit...and the hundredth visit. Sales and marketing strategies that compel your visitors to become customers include:

- Customer-friendly site design
- Push technologies
- Personalization of sites
- Customer service features
- Making each visit a new experience

Site Design

Web surfers really are surfers at heart. If they cannot find what they want quickly, they surf (jump) to the next site. Typically visitors click through three pages to find information, before they click to the next site. If you are the next site, great. If you are the three clicks, too bad. The goal is to make your site mimic the index of a book. A visitor should be able to reach the intended target within seconds or to be more precise, within three clicks, whether that target is product information, customer support, feedback, or the ordering screen. Some of the general rules of customer-friendly site design follow.

Place navigation options on the top or side of the Web page so customers can easily determine where they are at the site and how to navigate to other areas of the site. Provide these options consistently throughout your site.

Do not force visitors to register unless absolutely necessary. We have all visited sites where the company wanted us to register as soon as we arrived. Unless we had to gather information from that site, we would move on to the next site. Other sites require potential customers to register prior to shopping. Another barrier to entry. Forcing people to spend five minutes registering prior to purchasing a product may result in the potential customer becoming your competitor's customer.

Avoid using frames. Although Web designers cherish frames, consumers find them difficult and confusing (too many scroll bars and often complicated navigation). In addition, many search engines do not index sites that contain frames. Most of the top Internet sites do not use frames because of potential customer confusion and the lack of search capabilities.

Above all, use navigational elements, that lead your customer to the appropriate page as efficiently as possible. The Web's dominant sites feature simple and efficient design.

These rules apply, whether your site caters to business-to-consumer or business-to-business commerce. Information about Web demographics, geographics, and usage patterns, and the costs of developing Web sites is available from Cyberatlas at www.cyberatlas.com.

Push Technology

A few years ago, a company called PointCast offered software that allowed people to receive in real-time, news on stocks of interest, selected company information, and other news and mainstream media feeds. Users did not have to search the Internet for the information. All the users did was select categories of information. That information was automatically updated in real-time on Web browsers. With the vast amount of information available today, push technology provides an option to push information directly to your customer's browser. Why not provide your customers with this functionality? Several commerce software vendors, including Marimba, Microsoft, and Netscape, provide this functionality as part of their e-commerce solution.

With push technology, you can push marketing information, price lists, sales information, product updates and more. Consider how valuable this would be to your customers as they receive the latest updates, product tips, industry news, promotions, etc. This technology proves compelling even within the business-to-business market. Vendors can provide companies with new product descriptions and special promotions on items, without their customers having to search for the information.

Personalization

When you walk into a department store, you go to the children's section to shop for children's clothes, the teen's section to shop for trendy clothing, and the adult's section to shop for adult clothing. Personalizing your Web site means leading your customer to the appropriate information with minimal effort.

You should design your site to mimic the concept of the information desk or sales clerk. You are stationed at the front of a store. As customers enter, they ask you a series of questions. The first customer is shopping for a sweater. You ask "what size" and "is it for you?" You can see the customer, so you guess her age. Then you direct her either to the adult or the young adult area. Your customer may request a specific color, such as red. So you show her different red sweaters by a variety of designers. Then she asks you if you have a small sweater. You indicate that the sweater is in your stockroom and that you would be happy to wrap it for her. Most customers are motivated by price. Several sites include options for customers to indicate price range before any other responses are gathered. The site design process can prove complicated because you need to anticipate a typical transaction, and also account for any diversions from a typical transaction. The goal is to help lead your customers to the right place based on their needs. In a sense, you create an online salesperson.

Another navigational technique allows the customers to select the major categories or type of information that interests them. When they return to the site, they initially view that selected information. In several business-to-business e-commerce solutions, companies have set up sites so that when a vendor or buyer logs onto the site, based on their user name and password, they are immediately directed to products and services of interest.

Customer Service

After a customer makes a purchase, generate an e-mail message offering thanks and asking about the experience. If a customer is unhappy and replies to your message, you have a opportunity to correct the situation. Otherwise, you risk losing that customer. To offer a more personal touch, you might institute a policy of providing phone support in addition to e-mail-based support. Remember that it is much more efficient to keep an existing customer, than to attract a new customer.

Make your Site a New Experience

You want customers to return to your site to see what is new. Add elements to your site that compel your audience to visit your site frequently. Provide information that continues to attract them. Examples include product updates, tips and tricks, new promotions, etc. Several top sites provide their customers with news updates, onsite games, and other components that make shopping at their site interesting.

Tracking Purchasing Trends

Site analysis helps you determine what products or services your customers buy, those that they ignore, and the actual process of moving through the pages of your site. Several site analysis tools support your gathering and interpretation of facts. The great feature of these tools is immediate feedback. You can use them to determine exactly what products are doing well at any time. Specifically you should deploy a site analysis tool to track the following information:

- Number of items customers have purchased associated with promotions. This measure allows you to determine and track the effectiveness of your promotions. In addition, since you can place promotions on any one of your Web pages, you can track which pages users did not even access.

- Number of customer clicks on banner ads or featured-product ads.

- Number of products or services purchased online.

- Product, service, or other pages that were accessed infrequently. This infrequency may indicate that the information is too difficult to find. Typically, if a customer cannot find what they are looking for after three or four clicks, they move to another site.

- Product, service, or other pages that were accessed frequently. You also want to know what you are doing right, so that you can continue.

- Average time a customer remains at your site. If the duration of a visit is twenty seconds or less, the customer is not buying anything. Do you need to redesign your front end? Most sites function well as sales mechanisms when their pages exude simplicity and load (to become visible within the browser) quickly.

The technologist should not determine what information is valuable for selling goods over the Internet. Ask the product development, sales, marketing, and management teams what type of information is valuable to them for determining sales trends, and marketing trends. Exploit site analysis tools to your full advantage. The data imbedded in your site can become compelling information.

Marketing and Advertising Agencies

Now you know what to look for, but perhaps you lack the expertise or resources to focus on this effort. Several marketing companies can support your marketing efforts in this new global economy. Even if you decide to employ a marketing agency, do some preliminary work on the marketing strategies described in this book. When you talk to representatives of these companies, you possess the knowledge and background necessary to ask the appropriate questions. Most importantly, no one knows your business like you do. So you want to make sure that you are involved in developing your company's strategy.

Beware the marketing agency that hands you a template and then tells you that "this" is the way you need to market. This agency is probably not suitable for you. Yes, there are some standard methods of marketing. However, your business is unique and the company you select to support your efforts should develop creative ways of conveying that idea to your customers.

One marketing agency, Advanced Marketing Corporation, serves as an example of the type of support you can expect. Their service, Ultimate Promotion (UP), helps boost your Web site traffic. Their Traffic Building Package submits your Web site to more than 900 search engines and directories quarterly and submits your Web site to over 500 popular links pages. In addition, UP advertises your Web site with 5,000 displays of your banner ad. Each of these services generates visitors to your site. Your goal is to convert that visitor into a customer. Additional information describing the company is available at their Web site (www.theultimatepromotion.com).

Table 2.3 lists a variety of companies that can assist you with your marketing and advertising efforts. A complete list is available at www.adresource.com.

Table 2.3 Marketing and Advertising Agencies

Company Name	Internet Address	Profile
THINK New Ideas	www.thinkinc.com	A full-service marketing and communication company
Beyond Interactive	www.gobeyond.com	Features media planning, buying, and campaign management
Ad Up	www.ad-up.com	Offers site and banner design, as well as marketing consulting
Online & Multimedia Marketing Group	www.onlineadvertising.com	Offers e-mail marketing, Web development, and offline promotion
K2 Design	www.k2design.com	A full-service interactive firm
Eyescream	www.eyescream.com	Provides online media planning, placement, and strategies, and creative development for banner and advertising campaigns
USWeb/CKS	www.usWebcks.com	Helps clients define strategies and build their business
Deadlock Web Design	www.deadlock.com	Web site promotion and design

With the right marketing techniques the idea that "when I build it they will come" becomes a reality. The goal is to get as many people to your site as possible, keep them there to make a purchase, and repeat. Sales are the direct result of your marketing efforts. Think about it, even with traditional sales, marketing attracts the masses to the store. Then products, customer service, and other promotional efforts encourage those customers to return. Combine traditional selling methods with those offered online.

When you reach your goal of creating and maintaining a popular site, the result is increased sales and even increased revenue generation by selling advertising. So keep your eyes open and thoughts flowing. Use the techniques presented in this chapter to get the results you desire.

In the next chapter we focus on the globalization of your site. You are entering a worldwide marketplace, exploit it to your advantage. Globalization of your site ensures that you are not overlooking any potential buyers. Now that you have decided to jump on the e-commerce bandwagon, you might as well go all the way by globalizing your site.

3

Globalization

- Entering the new global marketplace
- Removing language barriers
- Creating the right site for the audience
- Distributing products worldwide
- Offering a 24 x 7 operation

A community's traditional market is determined by the number of consumers that can be reached effectively and efficiently in order to sell goods and services. Today's market goes beyond your hometown, city, and state. It has extended worldwide from the United States to Canada to Central and South America to Europe to Asia to Africa and everywhere in between. Global electronic commerce eliminates the geographic boundaries that restrain your trade, allowing even the smallest company to reach around the globe. E-commerce levels the playing field for both small and large businesses. It allows diverse businesses, small or large, urban or rural, well capitalized or on a small budget, to compete on the same-sized screen.

The world has not witnessed such a dramatic change in business since the Industrial Revolution. During that period of opportunity, companies that embraced the new methods of doing business rose to new heights. History repeats itself. If you had been involved with Henry Ford when he used the assembly line to build new cars, where would you be today? Think of e-commerce as that same moment in time. Those companies willing to embrace this technology and make the commitment will reap the benefits.

This chapter focuses on areas that prove critical for employing e-commerce for a global market. Specifically, the sections define the elements of global e-commerce, offer some considerations, and answer a variety of questions.

- Localization: What are the trends of e-commerce sales in different countries? Will you need to provide localization for some countries?

- Languages: Who are your current customers and how can you expand your markets to other countries? Do you need to have a multilingual site? What are the costs versus the advantages of providing multilingual sites?

- Demographics: Who is buying your services and products? You may think it is the technically-savvy Internet users. Think again. Based on the statistics and trends, your market ranges from the young to the elderly, as well as male and female, across many cultures. So make sure your site supports your target audience.

- Fulfillment and Distribution: How do you efficiently package and distribute these goods through the United States and throughout the global market place? Have you established and tested distribution channels?

- 24-Hour Access: Do you allow customers to purchase products and services 24 hours per day? If the marketplace is global, your e-commerce solution must be global.

The E-Commerce Global Market

The e-commerce marketplace primarily focuses on selling goods and services to people and companies located in the United States. As a country, the United States still represents the top market. However, the numbers are shifting. American markets are still moving upwards; however, other countries are recognizing the opportunities and jumping on board. They participate not only on the buying end as consumers, but also in the business-to-business and business-to-consumer markets.

Moving to the Internet to sell products and services requires selling globally. You can now run a 24 x 7 (24 hours by 7 days) business. Anyone with online access represents a potential customer. Often, consumers in other countries realize substantial savings when purchasing abroad. This opportunity has led to an enormous shift in buying power. It is important to consider several issues when addressing the global market. Market expansion and additional sales are part of this new sales model. However, they are not the only pieces of the puzzle. In order to increase the number of buyers from other countries, you must prepare for the shift.

Global Statistics

Consider the statistics that represent your current and future markets on the Internet. Begin with a look at current and future growth trends by geographic region in the United States and other countries worldwide.

An analysis from International Data Corp. estimates that in 1999, foreign firms will outpace the United States in e-commerce sales. The forecast shows that in 1999, United States spending will total $174 billion, while foreign firms will reach $305 billion. In addition, by 2003, companies will spend a total of $2.2 trillion, approximately $600 billion in the United States and $1.6 trillion by foreign companies. These figures reinforce the importance of the concept of a global marketplace.

The number of Internet users around the world is constantly growing. The Computer Industry Almanac (www.c-i-a.com) has reported that by the year 2000, 327 million people around the world will have Internet access. The top 15 countries will account for nearly 82% of these worldwide Internet users (including business, educational, and home Internet users). By the year 2000 there will be 25 countries where over 10% of the population will be Internet users.

With these types of trends, you will need to think about your business and how it addresses the world markets. Again, the idea is to move ahead of your competitors and to increase your market share. By globalizing your site, you have the opportunity to capture even more markets. Address the following considerations when designing your e-commerce site:

- What languages will you need to include in your site?
- Will you need to customize your site for each country?
- How do your products and services fit within the global economy?
- Will you need to provide different products and advertising for different countries?
- How will your business support the expansion to a global economy?
- Have you considered distribution channels, 24-hour access, import and export laws, taxes, and currencies?

If you choose not to translate or localize your site, but still want customers from countries outside of the United States, at least promote your site through advertisements, newsgroups, and cross-links. In most European countries for example, English is understood by a large percentage of the population. A report by Computer Industry Almanac estimates the following figures for the number of users online at the end of 1998. Only the top five countries are included in Table 3.1.

Table 3.1 Number of Internet Users

Country	Number of Internet Users
United States	88-92
Japan	13-14
UK	9.2-10.5
Germany	9.0-12
Canada	7.5-8.5

* All figures represent millions and are based on 1998 projections.

Global Trends

Take a look at some other current trends from the top countries online and the anticipated trends for the next few years. The reason they are important is that in business-to-consumer markets you need to compare the costs versus the benefits of localizing a site or providing additional language options. On the other hand, for business-to-business systems, you need to consider who are your trading partners. This consideration determines whether you should localize or provide additional languages on the site.

IDC Research has reported that Internet use in China is expected to surge to 9.4 million users by 2002, up from 1.4 million in 1997. China's Xinhua news agency announced in January of 1999 that there were 1.5 million Internet users in China at the end of 1998. More than 84 percent of these users were under the age of 35. China has 14,000 Internet Service Providers, with 3550 in Beijing, according to Xinhua. The Computer Industry Almanac notes that there were 1.58 million Chinese online at the end of 1998.

NOP Research estimates that 2.9 million French use the Internet (6 percent of the population). The Computer Industry Almanac estimates there were 2.58 million French online at the end of 1998. A Mediangles survey of 6,850 French citizens over age 15 conducted in May of 1998 led to an estimate of 2.5 million French citizens online. Whichever number is more accurate, one thing is for certain: when the 35 million French citizens who use Minitel are given Internet access (a plan in the works with the help of IBM) the actual number of French online will surge.

The German consulting firm GfK estimates that there are now approximately 8.4 million Germans with Internet access. More than half of these users are between the ages of 20 and 39.

According to a report by Osservatorio Internet Italia in December of 1998, 2.6 million Italians claim to have used the Internet in the previous month. In mid-1997 the number was estimated to be 1.5 million. Of the 2.6 million, 130,000 have purchased goods online, and 770,000 are daily users of the Internet. To compare, the Net user/Net buyer ratio in

Italy is 20:1; in the United States it is 3:1. The Computer Industry Almanac estimates that there were 2.14 million Italians online at the end of 1998.

Nikkei Market Access reported in October 1998 that there are 11.5 million people accessing the Internet in Japan. The Computer Industry Almanac estimates there were 9.75 million Japanese Internet users online at the end of 1998. DSA Group (www.dsasiagroup.com) reports the number is 14 million, with a forecast for 27 million set by Access Media. One thing is certain, the number of women on the Internet in Japan is growing. Studies by Nikkei Multimedia and DSA Group both report that 40 percent of the new Internet users in Japan are women.

Almost 2.25 million Spaniards (6.6 percent of the population) have Internet access, according to a study by the Spanish Internet Users Association (AUI). It predicts that the number will rise to 8.74 million by 2001. The Computer Industry Almanac estimated the Spanish online population to be 1.98 million at the end of 1998. Clearly these trends are headed in the right direction for any company considering a global presence.

Big Business Goes Global

With business-to-business commerce, international companies are communicating via the Internet and embracing e-commerce as a huge new opportunity for their business. If you are currently selling goods or are considering selling goods to other markets, languages and localizing your site may be the keys. E-commerce worldwide has taken several forms as companies create malls for suppliers, manufacture and distribute goods, and sell products to retail stores throughout the world. The following companies and their strategies represent the way organizations outside the United States are taking advantage of e-commerce to communicate and to sell with business-to-business transactions.

Best-of-Italy built a mall site to provide connectivity, services and visibility to suppliers. Their strategy was to enable new suppliers to easily rent space on their site so that they did not have to create their own. The Best-of-Italy mall also had to enable secure online payments.

For individual suppliers participating in the Best-of-Italy online mall, the benefits of the best of Italy site are clear. They establish an online presence without investing in their own solution. For Best-of-Italy, the benefits are equally clear. The IBM solution relieves them of the burden of developing and innovating the site's technology. And for customers, every aspect of the transaction is handled by Best-of-Italy, so they get the low prices and reliability they are accustomed to as Best-of-Italy customers.

Lister Petter, a manufacturer of industrial diesel engines, runs its worldwide manufacturing and distribution from Gloucestshire, England. The parts distribution side of the business needed to upgrade its paper-based system which was slow and restricted the company's ability to respond to marketplace trends. Lister Petter needed a tool that could pull together information from different systems across the world.

The new system is easy to use and requires little or no investment from Lister Petter's network of distributors. It makes the ordering process more efficient. IBM helped Lister Petter achieve these goals with minimal expenditure, keeping this small company competitive.

Office Depot of Mexico wanted to expand its established business, a catalog service and phone ordering system, through the Internet. Office Depot of Mexico has experienced improved customer satisfaction, better inventory management, and a sales increase since implementing its e-commerce solution. With the help of IBM's expertise and e-commerce products, they established and continue to build on an infrastructure that emphasizes scalability and the integration of emerging technologies. More information about these and other companies can be found at the IBM Web site.

Language Considerations

Language considerations address the languages people speak instead of any geographic concerns (the specific countries that they inhabit). It is important to distinguish between the two when considering the additional capabilities you might add to your e-commerce site, such as multilingual support or localization. Some countries that your site targets may only require a single language (with English as an alternative). Businesses in most countries outside of the United States provide sites that contain both their native language and English. Other multilingual countries, such as the United States or Switzerland, may require support for several major languages. Remember to consider the native language of your audience, whether you target a single region or a single population. Globalization requires spending some time researching you markets and determining the best plan of action.

Figure 3.1 Growth of the Customer Base

YEAR	DISTANCE TO STORE	SIZE OF CUSTOMER BASE
1800's	1/2 mile / Walk to stores	
1900's	10 Miles / Drive to store	
2000's	20 Feet / Walk to home office	

Making a Choice

The driving force to include multilingual support differs between commerce sites providing business-to-consumer solutions and those providing business-to-business solutions. For business-to-consumer endeavors, the decision to provide multilingual support is driven primarily by the top languages spoken worldwide and the top countries currently online. For business-to-business endeavors, the approach, not driven by total online populations, instead is driven by the demands of the businesses you are connecting. For example, if an American company's primary customer is a Japanese company, it may be a competitive advantage to provide a Japanese (translated) site so the two companies can easily communicate. However, in most business-to-consumer commerce solutions, it may not necessarily prove a competitive advantage to maintain a Japanese site. So when reviewing the option to provide a multilingual site, approach your decision with a focus on whether you primarily link to consumers or to other businesses.

Table 3.2 contains some of the latest statistics on the number of language populations online. The populations should provide you with a starting point in determining the languages you want to support online. These figures were compiled by a company called Global Reach and are available as a complete listing at www.euromktg.com.

Table 3.2 Projected Online Populations

Language	1998*	2000* (projected)
English	107.2	160
Japanese	14.4	23
Spanish	14.2	Not listed
German	13.9	25
French	8.3	16
Chinese (Mandarin)	6.4	Not listed

* All figures represent millions.

This list contains only the highest language counts with Internet access worldwide. If you view this in terms of growth, the Japanese, French and German languages are expected to increase by 40% in the next year or so.

Figure 3.2 Language of Internet Users

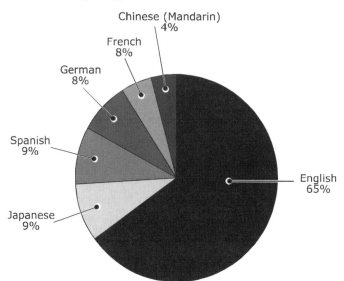

Many prominent companies have deployed and continue to support multilingual sites. They realize that their marketplace is global and that they must address their foreign customers by catering to different languages, cultures and time zones. Representative companies include Barnes and Noble, Lands' End, and Dell.

Barnes and Noble, the United States bookselling giant, looked for help when it wanted to sell overseas, partnering with German media giant Bertelsmann to run its online ventures in Europe. The joint operations allow Barnesandnoble.com to offer localized content, distribution and operations in Germany, France, the United Kingdom, the Netherlands, and Spain. Unlike some other companies, the international sites are branded separately as BOL, which stands for Bertelsmann Online. But all the sites provide links to the United States' site.

Lands' End, a catalog-driven global organization, uses its own overseas operations to set up online overseas. The company has separate Web sites for Germany, Japan, and the United Kingdom, each with its own localized address. The German and UK sites provide company information, allow consumers to request catalogs, and offer e-mail customer service. The UK site also gives information about physical stores in that country. To date, neither site allows online transactions, instead referring customers to the United States site. The Japanese site is transactional, allowing consumers to order catalog items and overstocked items. It also offers some features specific to the Japanese market, such as a dictionary of specialized words and phrases used in the clothing industry, and links to entertainment articles.

Dell has been doing business internationally and online for years, so it is no surprise that the company has specialized sites for dozens of countries. And its products, unlike books or clothes, need to meet different technical specifications depending on the country. Power supplies and encryption software are just a few of the components that vary by country, as do tariffs. International customers are unable to buy from the main American site; indeed, the order page requires consumers to state that they will not export the systems out of the country. But Dell's main home page includes options that take a user to a specialized section of the site with local content, currency, and language.

International Marketing Firms

If your company does not currently do business globally, you should consider working with a marketing company that guides you in the right direction. Some of these companies include Global Reach, International Data Corporation, and BlueSky International Marketing.

Global Reach (www.euromktg.com) assures your company a comprehensive approach to building traffic to your Web site in all online countries where there is a critical mass. Global Reach addresses every means that people use to locate a Web site in any particular country. The key is their network of marketing specialists based in these countries who know their local Internet market intimately. Global Reach ensures that people from many countries visit your Web site.

International Data's Worldwide Services research program (www.idc.com) offers a comprehensive view of service industry trends and identifies opportunities across major geographic regions. In this program, IDC assesses competitive dynamics and global strategies for revenue growth in the IT services industry. Then they position the IT services organizations in the global marketplace. They also provide detailed market forecasts and trends by region, by service category, and by vertical industry.

BlueSky International Marketing (www.bluesky-inc.com) represents European Internet marketing, research, and communications specialists. They help you find your competitive edge, country by country, market by market, and they deliver detailed and actionable strategic plans. They also provide the market knowledge you and your team need before you enter the Internet in foreign territory. In addition to these three companies, other strategic marketing firms can be found at www.adresource.com.

Translation: Translator Service versus Software

Some companies have set up Web sites that allow a visitor, whether a consumer or a business representative, to select the specific county that the site targets. Once the country is selected, the textual information is presented in a native language, such as German or Chinese.

Other e-commerce sites have approached this target by orienting to a specific language. When visitors access the main site, they select their choice of language. After selection, the information is presented in the chosen language via an alternate site.

Depending on the type of data on the site and the scale of the business, several options are available for preparing your site for the international marketplace. These options range from implementing software that translates Web sites on-the-fly to employing a translator. Take a look at how these options may be used considering your business needs and budget.

A traditional method for translating a site relies on the services of professional translators. These translators, as employees of translator service companies, translate the static elements of your Web sites into each of the languages you select. Static elements may include product information, customer service information, Web page titles and sub-titles, policies, etc. These companies typically price their services on a project-by-project basis or according to the amount of content translated. Consider dominant vendors when choosing a translator service company. Berlitz Global Business of Princeton, New Jersey (www.berlitz.com), translates sites into 200 different languages. Additional services include project planning and analysis, resource allocation, glossary development, production of audio, video, and graphical elements, updating links, integrating localized files, cross-cultural evaluation, testing, and validation. Bowne Global Solutions of Los Angeles, California (www.bowne.com), offers Web site localization services to manually recreate a site in 40 languages. They also perform real-time machine Web translation. International Communications of Framingham, Massachusetts (www.dlc.com), recently merged with Direct Language Communications, translates Web sites into 50 languages. Services include localization of products, creation and design of multilingual Web sites and electronic commerce capabilities, software testing, international branding, print and electronic publishing, typesetting, and digital color printing. Consider the scope of your project and how well the services offered match your needs before selecting a company.

Keep the following rules in mind when you select a translator service company. Choose a translator who understands your product or service. The translator should have the technical credentials to accomplish the work accurately. Choose a translator who speaks the language of your target area as a native. Choose a translator who understands your message, and knows how to write copy. The translator should be able to incorporate local dialect, customs, and cultural norms into the work. Choose a translator who can be employed for additional projects later. The process of finding a quality translator is tiresome enough that you do not want to frequently repeat it. Find translators that you trust and then, trust them.

If your site contains a combination of static text and dynamic text (information that is constantly changing), such as product updates, FAQs, new releases, etc., use a combination of translation services and on-the-fly translation programs. In addition you could use the on-the-fly translation method to deliver languages which are not cost-efficient to use a translator.

In some cases, return on investment is not easily determined. In other cases, the return of small investments includes compelling benefits. We recently gave a presentation to a major department store company and asked the question, "How many Spanish-speaking customers do you have?" Based on their response we wondered aloud why their site did not include any information in Spanish. They responded that the translation of their site into Spanish was so technically challenging that it proved cost-prohibitive. Thirty seconds later we located a package of site translation software that could immediately translate a site with 90% accuracy. The company representatives were speechless. Lesson? Do not let information technology drive your company's solutions. Let the business drive the technology.

There are several vendors that provide translation software that translates a Web page within seconds. This software is a great option for those businesses that do not have the resources to translate an entire dynamic e-commerce site. Typically, the translations are approximately 80-90% accurate. This measure is adequate for a site if you want to allow your potential customers to at least see the information in their native language. Some of these software programs even provide translations into fourteen different languages.

Several companies now provide services catering specifically to the global market. Some of these services include localizing your sites, multilingual site translations and software translations. As a starting point, consider some of these companies that jump start your move into the global market place. Systran (www.systransoft.com) offers translation software that provides the following translations at 3,700 words/minute (in either direction) from English to French, German, Spanish, Portuguese, Italian, Japanese, Chinese, Russian, and also Korean. Transcend (www.translc.com) also generates translations on-the-fly. They have free software that you can test for a limited period. WorldPoint (www.worldpoint.com) of Honolulu, Hawaii sells a software suite called Passport. This suite of software-based content tools enable a company to develop scalable, multilingual sites in up to 75 languages using a network of 6,000 human translators worldwide. Passport Pro is for 10 users or fewer. Passport Enterprise is for corporate intranets. Passport CSP brings it to the Internet Service Provider (ISP) level.

Demographics

Who uses the Web? Who buys via the Web? According to a recent Pew Research Center study, women account for more than half of new Internet users. By the end of 1999, Forrester Research Inc. predicts that 32% of African-American, 43% of Hispanic, and 67% of Asian-American households in the United States will be online. These proportions compare with 39% of Anglo-American households expected to be online by the end of the year.

Internet access and e-commerce participation is not limited to the younger generations. According to Forrester Research, those aged 55 to 64 comprise some 22% of online households today. And they will reach 40% by 2003. Forrester also suggests that older citizens appeal to Web businesses, ranging from sites selling flowers to automotive retailers, because household incomes are typically above $60,000.

These are your markets. Web usage and Internet purchases continue to rise across every corner of society. The technically savvy have been joined by relatively inexperienced audiences who intend to experience the Internet on their own terms. What does this mean to you as a merchant? Be prepared for your audience.

Targeting Your Audience

Given the diversity of audiences, ensure that your site design targets the consumer group that most benefits your business. Some of the Internet's best targeted sites, such as www.bluemountain.com and www.women.com, focus on personalization and have performed extremely well in attracting ethic groups, women, and seniors. One of the key features of these sites is simplicity. If you are trying to attract customers that are not technically savvy, it is important that the site includes easy-to-access information and simple navigational elements. Plug-ins, elaborate frames, and conflicting color schemes should be avoided. The worst offenders tend to be sites that require multiple downloads of plug-ins prior to offering complete functionality. Keep the interface simple. This includes clear categorization and appropriate color schemes. Change the techno-lingo to easy to understand wording. Focus on the specific demands of your audience.

An important way to support the sale of products or services at your Web site is to include a toll-free phone number. Allow those people that are concerned about Internet security to place their orders (and use their credit cards) with a toll-free phone call. You are more likely to become a victim of fraud by handing some waiter your credit card after dinner, than you are by charging a product online. But some Web site visitors remain skeptical. Other techniques to make consumers feel secure are to incorporate a policy statement and a privacy statement into your site, and to specify that your site is verified by a third-party certificate authority, such as Verisign. Now sites are also posting the Better Business Bureau logo to show that they belong to the organization. As more people experience online shopping, security fears will subside. But for now, do not miss a sale just because your customer is tentative.

Marketing Companies That Use Demographics

Consider the following demographic issues:

- Who influences a customer's decisions? Consider kids, spouses, friends, celebrity endorsers.
- Who will pay more for a brand name?
- What variables distinguish heavy, medium, and light users?
- Do demographic, lifestyle, and brand user/volume profiles of current and target markets differ?

Several companies now help you target a specific audience. They may be able to help your company ensure that you have not excluded any portion of your potential market. Look for complete descriptions of these companies at www.adresource.com. One prominent company, Mediamark Research (www.mediamark.com), offers both research and directed marketing to answer the questions posed above, ensuring that your e-commerce effort is successful.

Getting Product to Customers

Distribution is a key component to carefully plan out before going online. Once you go live, you need to have your distribution channels in place to ensure products are shipped to customers in a timely and cost-efficient manner. One problem that companies have during start-up is that their distributions channels are not in place when they go online. As a result, customers experience delays in product delivery, and merchants experience higher shipping costs. Again, once you hit the Web you have a worldwide audience. Make sure you have close communications with your suppliers and your distributors to handle all orders.

Product Size and Weight

Do not force a square peg into a round hole. Determine very early in the planning process whether your products can be reasonably shipped. If the cost to consumers are significantly higher to purchase online, they may prefer to go down the street or to the mall to buy. Typically, smaller items are much more cost-efficient to ship. In addition, if your products are of similar size, the packaging issues are less complicated, with multiple products fitting in the same size containers. Also, consider the efficiencies of shipping from individual stores versus from a central distribution center.

Inventory

Another area which is important to consider is inventory requirements. When moving online, it is important to know how much inventory is needed to get started. In addition, if inventory is low, ensure that you know exactly how long it might take to replenish your stock. Initially these figures may prove difficult to determine, however you will eventually gain a better sense of how to meet the requirements.

Distribution

Distribution of products after the order is placed and fulfilled begins with determining the best method of shipping products to customers in the most timely and cost-efficient manner. For larger companies that currently sell goods via catalogs or direct marketing, current distribution channels are used as the method for shipping products to customers. Those companies that are not currently shipping to customers need to determine the best method to ship goods.

Companies such as FedEx, UPS, and the USPS have special pricing for merchants that are moving online to sell goods on the Internet. When reviewing these shipping methods, determine shipping/handling costs, freight costs, and the cost for overnight, next day, or regular shipping. In addition, determine how much of these costs can be paid for by the customer. If shipping costs are too high, some customers may decide to shop elsewhere. Make sure the cost is competitive in the marketplace. If the costs of shipping are too high due to the size or weight of a product, this may be the wrong product to sell online.

Companies with sites in different parts of the United States and worldwide can use current distribution channels. For example, ship from local stores. Other savings can be generated by wiring your suppliers. The best method is to set up agreements with the suppliers of the products being sold so that when a customer orders a product, these products are sent directly from the supplier to the customer. The benefits are two-fold: reduced shipping costs and faster delivery. At least in terms of distribution, cut yourself out as the middleman.

Fulfillment Companies

Another option is to use fulfillment services to take care of the ordering, replenishment, and customer service aspect of the business. Although this method may be a bit more expensive, it also has its advantages. Instead of thinking about filling and distributing orders, you can focus on your business, marketing, sales, and on gaining customers.

Fulfillment companies provide a variety of different services. National Fulfillment Services offers order processing, call center, customer service, financial and marketing reporting, and distribution. DupliSoft Inc., specializes in services such as replication (copying), packaging and distribution of software and related technology (i.e. modems, digital cameras, external storage drives, and even digital telephones). Fill It Inc. provides quality control, technical reference service, customer reporting, computer services, and software and consulting services. In addition, they offer an option to link from your page to their e-commerce engine. When customers use a shopping cart to purchase products, the orders are processed and an invoice is sent to their warehouse, where the order is filled and shipped. Other fulfillment companies include Equire, Direct Marketing Resources, Quality Fulfillment Services, Inc., and SalesLink. More fulfillment and distribution companies can be found online by searching at Yahoo! or any other major search engine.

Keep these checklist items in mind when selecting a fulfillment company. What products comprise your inventory? How large is your current and potential inventory? How much budget can you provide for fulfillment services? What types of services do you expect from the fulfillment company? Are your business systems compatible with the fulfillment company you selected?

24-Hour Access

Since your site can be accessed by anyone at anytime in any part of the world, change your focus to 24 x 7 (24 hours a day, 7 days a week). When designing the site features and functionality, make sure that customers can purchase products at anytime, and even more critically, that online customer service can be delivered online: directly, precisely, and automatically.

For companies that do not currently provide 24-hour customer service, you should consider this option. If your market is primarily comprised of American visitors, at least support your customers by extending support hours to cover time zones from east coast to west coast. Another method of accommodating additional support requests is to allow customers to type in the specific information needed online so you receive the information immediately and can call them back the following day.

For international sites, have bilingual service personnel available to answer calls from countries abroad. While this practice may prove a bit more expensive, if you are an American company with a large Spanish or Japanese speaking customer base, it might be well worth the expense.

Web Access Trends

A majority of e-commerce consumers access the Internet via a personal computer. However, there are several trends emerging that alter the methods of people accessing the Internet and shopping online. These alternative methods will increase the audiences that have access to shopping in an online community. Just imagine, for example, sitting in front of a television and visiting online virtual stores. The potential audience is overwhelming. Let's peek at some of these alternative methods of Internet access to gain a glimpse of the not too distant future.

According to IntelliQuest Research, more than 83 million adults, 40% of the American population age 16 or older, are accessing the Internet. Of these users, 3.7 million use a portable computer to access the Internet and 3.1 million use a television set-top box or WebTV. They also forecast that by the year 2000, the market for set-top boxes and information appliances will exceed two billion dollars.

As alternative methods of accessing the Internet flourish, especially WebTV, buying online will resemble buying on QVC or the Shopping Network. Although it is not mainstream yet, who knows what the future holds? Be prepared. Keep your eyes and ears open.

Electronic commerce allows buyers from every corner of the globe to walk past your storefront. Throw your doors open, and keep the doors open throughout the day. Be prepared for customers as they enter your store. Recognize their needs. Speak their language. Address their dialects, customs, and culture with appropriate site design. Simplify the buying process. Deliver your product or service on time and inexpensively. Thank them for their patronage.

4

Company Profiles

- Learn about successful companies
- Generate advertising revenues
- Service your customers
- Save money by managing your supply chain
- Sell to consumers via the Internet
- Sell to other companies via the Internet

Anyone who has paid attention to the e-commerce world, even from its periphery, understands the dynamics of its dominant companies: Amazon.com, America Online, eBay, and Yahoo!. Amazon.com stands at the summit of online retailing with its capitalization and its ability to anticipate future trends. The company has created a strong brand and a growing community of customers; there are currently 6.2 million. The media has even picked up on the term "amazoned," referring to the tactics and techniques Amazon.com uses to address specific markets.

America Online, with a membership of 16 million (and growing), harnesses a lowbrow approach featuring corny television advertisements and mass mailings of floppy disks and compact disks to grow the company. Other online companies that want to generate traffic to their sites often sign marketing deals with America Online. With its recent acquisitions of Netscape and Mirabilis, America Online has become a dominant force in the e-commerce arena.

The auction model may one day dominate online transactions. eBay, the premier auction site on the Web, facilitates the buying and selling of merchandise and collectibles, without having to fulfill the orders. What a compelling business model. The company now has more than 2.1 million registered users and a recent market capitalization in excess of $17 billion.

The last of the four cornerstone companies is Yahoo!. This Internet portal has managed to combine the advantages of being first and being best. According to the research company Media Metrix, www.yahoo.com was the most visited address in January of 1999 with over 29 million unique visits. Driven by the sale of banner ads and sponsorships, Yahoo! is also one of only a few Internet companies that actually turns a profit.

With as much media attention as these four giants receive, it is easy to overlook the wide variety of companies that inhabit and prosper in the e-commerce world. Companies with an electronic commerce presence can be categorized by their business models as they focus on advertising, servicing customers, managing the supply chain, selling business-to-business, selling business-to-consumer, and facilitating e-commerce for other organizations. By considering the different approaches to electronic commerce, the advantages to each approach, and several representative companies, you may begin to appreciate the diversity of business models and some alternatives for improving your own business efforts.

Advertising

Some companies may not participate in the electronic commerce arena. Their product or service may not lend itself to online delivery, or even electronic transactions. Perhaps their relationships with suppliers and buyers are fragile and complicated. Company executives may not want to cannibalize their existing distribution channels, whether they emphasize storefronts across the country, catalog sales, or door-to-door salesmen. A company may not employ the appropriate skills or knowledge to effectively develop an online solution. Even if a company ignores the benefits of electronic commerce, the Internet offers myriad advertising opportunities.

Advertising on the Internet, whether via e-mail or the Web, can generate returns that far exceed the effort and expense required to develop an effective online presence. Online advertising allows companies to market their products and services to targeted audiences. Determine the characteristics of your typical (or potential) customers. Are they separated by age, geographic region, native language, sex, level of literacy, interest, vocation, or wealth? Can you attract visitors to your Web site by addressing these differences? Advertise your products and services with an eye for the demands of your audience.

If you rely on toll-free phone calls, the United States Postal Service, or storefront transactions to sell your product and business, consider an online presence. Advertising online generates visitors and customers that might not otherwise be aware of your offerings.

Another advantage to advertising online is in your ability to gather customer information from your Web site. Collect this information by offering services online in exchange for registration or by responding to e-mail requests for service. Harvest the resulting information to learn more about your customers and their needs. Use this information to tweak your existing marketing materials or to create new marketing campaigns.

As your advertising efforts improve and your online presence gains popularity, begin the transition to electronic commerce. Consider the advantages of providing online customer service. Consider selling your products and services online.

Funjet

As travel services continue to spring up across the Internet, a variety of business models are being tested. Travel agents, airlines, ticket brokers, resorts, and automobile rental agencies all clamor for the attention of travelers. One vacation vendor, Funjet (www.funjet.com), has managed to create a compelling online presence without undermining the traditional role of its travel agents and without delivering a product electronically. Funjet, whose parent company is the Mark Travel Corporation, uses its Web site to advertise its charter-based vacations, to describe a variety of destinations, and to allow potential customers to plan and price their trips. Funjet vacations deliver tourists to Mexico, Las Vegas, Florida, California, Reno, Hawaii, Europe, the Caribbean, and to a variety of ski destinations. The site features "Hot Deals" and profiles of many hotels and vacation properties. The most interesting areas of the Web site, and what separates Funjet from its competition, are real-time planning and pricing mechanisms. Instead of waiting for a travel agent to respond to your specific requests, look again, and finally, complete the research process while you wait on the phone, Funjet allows you to build your own trip in a single online visit. With links to charter flights, hotel packages, transfers, car rental information, the site allows customers to select a time period, a destination, and a hotel, to add options, and to calculate a total price for the vacation. A travel agent then completes the transaction. Even with the restriction of their business model and their relationship to suppliers and buyers, Funjet still manages to employ Internet technology to add value to their advertising efforts.

Dallas Morning News

Another industry that continues to flourish on the Internet is electronic publishing of media that is also available via newsstands, home delivery, or the postal service. The Dallas Morning News (www.dallasnews.com) represents one example of the advertising power of the Internet. As the only major newspaper serving the Dallas metropolitan area, the Dallas Morning News offers a complete version of the daily edition free to online visitors. The online edition includes world, national, and local news, editorials, obituaries, weather reports, and sections covering education, lifestyles, religion, science, and technology. The sports section, recognized as one of the top ten in the United States, is included. Business and political news appears online, as well as an entertainment section. All the classified advertisements (employment, automobile, residential, commercial, and shopping-oriented) that appear in the printed version also appear in the online version. Web exclusive features include crossword puzzles, horoscopes, comics, and a variety of discussion forums.

The common problem facing the traditional publishers is how to harness the Web without jeopardizing current distribution channels and without alienating current customers. By using the Web primarily to attract potential customers, the Dallas Morning News secures additional subscriptions and advertising revenue, and increases the size of their market from the state of Texas to the entire wired globe. As viewership grows, the Dallas Morning News will be able to transition from an advertising medium to an online revenue generating medium.

E-Preneur.com

When a product or service can not be delivered via the Internet, some organizations rely on the Web solely for advertising. Advertising via the Web raises brand awareness, answers the questions of potential customers, and establishes an additional point of contact. E-preneur.com (www.e-preneur.com) represents a service organization that uses the Web on all three of these fronts to advertise and explain legal services, accounting services, and entrepreneur services. E-preneur.com's legal services include sales, purchase, or formation of businesses, Internet and e-commerce law, Web site development agreements, linking agreements, and multimedia law. E-preneur.com's accounting services include e-commerce business development, business plan preparation, tax advice, and tax return preparation. E-preneur.com's entrepeneur services include e-commerce architecture and consulting, Web site development, Microsoft Site Server deployment, and software/hardware support.

Wouldn't it be great if your attorney, CPA, and e-commerce architect were all on the same page? At E-preneur.com those skills just happen to reside in one person. By combining, legal, accounting, and technical skills J. Jeffery Johnston uses the Internet to advertise and explain his approach to e-commerce solution development. The E-pre-neur Web site explains each type of service, contains links to other resources, and includes contact numbers to request more information.

Servicing Customers

You have just completed a sales transaction. One more unit shipped. Should you turn all of your attention to finding the next customer? Not yet. An important component of the electronic commerce effort is realizing how much easier it is to keep an existing customer than to gain a new customer. How do you add value to a transaction? How do your customers communicate with distributors, service providers, and manufacturers? What do you offer your customers in the way of support materials? How do you help your customers solve problems? Customer service. Customer service. And customer service.

Servicing customers online offers many advantages, either in conjunction with your traditional customer service effort or as your main effort. The most compelling feature is the availability of customer services 24 hours a day, 7 days a week. Your automated customer service components never sleep. Customer service online, particularly frequently asked questions (FAQ) and targeted e-mail, offers immediate response to customer problems. The online experience can be tailored to each customer, driven by their typical needs or by their recent visits. Elements of your online presence can be immediately updated as you determine it necessary. Updating involves simply posting new information to a server. In addition, after absorbing the costs of establishing online customer service, development costs decline. Focus your design efforts on replacing existing customer service efforts (to reduce your overall costs) and adding value to your customer's shopping and buying experience.

TXU

TXU, formerly known as Texas Utilities Company (www.txu.com), based in Dallas, Texas, is an investor-owned company providing electric and natural gas services, energy marketing, telecommunications, and energy-related services domestically and internationally. TXU's current challenge involves harnessing the power of electronic commerce to improve the competitive capabilities of their multi-faceted utility. The professionals of TXU address this challenge by focusing on adding value to their residential and business customers with creative e-commerce solutions.

Within their Web site, TXU provides a variety of value-added services to residential customers. Home Energy Audit generates a free online profile of a customer's energy use, including tips on managing energy more efficiently. HEET (Home Energy Efficiency Tour) is a virtual tour of a home that teaches techniques for more efficient heating and cooling, insulation, lighting, etc. TXU's Web site also features E-Choice, a system that provides expert advice about replacing a heating or air conditioning system and referrals, if requested. In order to ask questions or express concerns, residents with access to the TXU Web site use e-mail to contact energy experts. These features represent only a few of the many ways that TXU uses their Web site to provide exceptional customer service.

TXU also supports the performance of their commercial customers, especially in the area of improved billing practices. By implementing electronic data interchange (EDI), the computer-to-computer exchange of documents in standardized, electronic formats, TXU offers paperless billing. This method of billing saves between $1.30 and $5.50 per bill by eliminating the time and labor required to process and pay bills manually. EDI represents a business strategy that deploys a technical solution to improve business relationships and to meet target objectives. It also forms the backbone of many business-to-business electronic commerce solutions. By gaining an understanding of how to capitalize on the potential of electronic commerce, managers are able to create compelling value-added services for their customers without the luxury of delivering products electronically.

Canon U.S.A., Inc.

Canon U.S.A., Inc., a division of Canon (www.canon.com), offers a variety of customer service features within its extensive Web site. In addition to selling accessories for their main line of products, Canon U.S.A., Inc. provides features such as frequently asked questions (FAQ), general and technical inquiry responses, information on toner cartridge recycling, replacement operator's manuals, and product information. If you have a technical question about a Canon product, the FAQ provides solutions for most common problems. Canon employs a large staff of professionals to answer questions regarding the use and operation of any Canon product. Printer owners can mail used toner cartridges to Canon for recycling, and can request boxes and shipping labels at the Canon Web site. Product owners can also request hard copy feature and specification lists for a variety of products.

In addition, the company also offers software downloads for Canon brand products sold in the United States. These downloads include informational guides, hardware drivers, technical field bulletins, and other software support products for printers, copiers, facsimile systems, scanners, digital cameras, personal computers, visual communication systems, and device management systems. By offering this variety of customer support features via its Web site, Canon adds real value to a sales experience that traditionally only occurs in a storefront setting.

Internal Revenue Service

The Internal Revenue Service (IRS) enhances its reputation across the United States by reducing the complexity and harshness of the taxpayer experience. With its Web site at www.irs.gov, the IRS delivers customer service to individual taxpayers, to small businesses, and to tax-exempt organizations. The IRS site, presented as an online newsletter called

The Digital Daily, incorporates tax statistics, tax information, electronic services, tax regulations, forms and publications, and e-mail support. Tax statistics available from the IRS contain information about the financial composition of individuals, business taxpayers, and tax exempt organizations. Tax information includes rate tables, terms, event calendars, state and local news, IRS news and notices, bulletins, and frequently asked questions. Electronic services incorporated into the site allow visitors to file taxes, pay taxes, and collect refunds electronically, search for employment opportunities, make comments, and even pose questions to an expert system. The IRS Web site also allows you to download most of the tax forms, instructions, and publications in several popular file formats. E-mail support delivers to a specified address important upcoming tax dates, listings of new information on the IRS Web site, recently added tax forms and publications, IRS news releases, and special IRS announcements. By providing such a variety of services, the Internal Revenue Service takes some of the sting out of the taxation process.

Managing the Supply Chain

Automating the supply chain is one of the loftiest goals that a business can attempt to reach. The ideal is to create a business system that functions automatically wherever possible, and reasonable. More typically, however, companies are content to improve their processes incrementally, automating where practical. Since most businesses are defined by the strength and coherence of their relationships to suppliers and buyers, improvements in supply change management can greatly improve a company's productivity. What are the components of your current supply chain? How does your company streamline the transactions between buyers and sellers? What changes could be made, using online technology, to enhance performance?

The benefits derived from effective supply chain management compel companies to contribute even more resources to the management effort. Such benefits include reduced expenses. Less time is spent on pushing paperwork, phone calls, faxes, and on tracking all of this information. For example, the typical purchase order costs between $75 and $125 to process manually. With the improved automation that e-commerce offers, that cost can be reduced to about $3. Bills can be customized for each client. Because many of a company's processes are automated, errors are naturally reduced (and often eliminated). The concept here involves creating a process that performs correctly, and repeating that process. On the inventory management side, an online effort can result in just-in-time inventory, reducing storage and handling costs, and if you manage clients' inventory, you can lock out the competition. All of these improvements are available within the supply chain, without even having to address the sales side of the equation.

webMethods

One strategy for improving supply chain management involves implementing an XML-based back office system for applications such as supply chain management, automated procurement, shipping and logistics, and aggregation of business intelligence. XML (eXtensible Markup Language), the universal language for structured data on the Web can drive cost-efficient, reliable, and open means of communicating between business partners and customers. Companies can leverage existing investments in virtual channel sales, marketing, customer service, and purchasing systems by replacing current and proposed electronic data interface (EDI) systems with XML. Customers learn that XML is easier to deploy, easier to understand, and can be adopted much more rapidly than traditional EDI solutions.

webMethods (www.webmethods.com), a company focused on delivering XML-driven products and services, enables enterprises to forge automated links with customers, suppliers, and partners quickly, easily, and cost effectively. Their B2B products enable companies to leverage investments in existing Web sites, applications, and local data sources, reducing the time and complexity of implementing true business-to-business e-commerce compared to technologies such as CORBA and traditional EDI.

The resulting benefits of this XML deployment are compelling. Clients gain tight intercompany integration, linked Web sites, ERP systems, databases, and EDI systems, improved customer service, leveraged current IT investments, an extensible, scalable solution path for their business partners (with no purchasing required), and a low cost-of-deployment. Implementation of an XML-based solution can be accomplished in weeks, instead of months or years.

The National Transportation Exchange

By allowing companies to reduce the expense associated with their supply chain, some facilitating companies exploit the Internet very successfully. For example, the National Transportation Exchange (NTE) at www.nte.net, a Chicago-area company, has created an Internet-based exchange for empty truck space, promising to slash $15 to $20 billion a year in distribution and transaction costs from the $400 billion trucking industry. The company's president, Greg Rocque, and his staff have enlisted 350 members so far, and expect to handle approximately 150,000 loads during 1999. Members pay a startup fee that varies according to the amount of deployment work that NTE engineer's must perform to mesh a member's databases with NTE's. NTE also collects a small fee for every match they make. Trucking company dispatchers can feed into a real-time network from their headquarters without having to dial into proprietary database systems. NTE helps trucking companies in two ways: by enabling them to calculate their profit margins on every transaction and by filling most back-haul (return trip) lanes, thereby removing inefficiencies from the transportation system. While other organizations offer Internet-based and bulletin board-based logistics, NTE's proprietary calculations fill a genuine national need.

BOC Gases

BOC Gases, a division of British Oxygen Company, is a global provider of gas products to the food processing, healthcare, biotechnology, and electronics industry. The company decided that it could reduce its clients' administrative burden associated with gas procurement by serving as a clearinghouse of industrial gases. It would supply the majority of gases and track the purchase and delivery of gases from third-party vendors for each customer. The key to addressing the technical issues of this proposal was locating a solution vendor that offered software development, server operations, and around-the-clock support. BOC Gases, working with IBM Global Services, created an Internet-based application that made it easier for customers to reorder gas products. IBM Global Services coded the application, which dynamically configures itself to a customer's account profile, including the particular product mix, contract prices, and payment method. The first major customer that used this system was able to consolidate nearly all of its purchases to BOC Gases, thereby reducing its 14 gas suppliers to only one. In addition to providing benefits to clients, the Internet-based system helps BOC Gases increase revenues, simplify account management, reduce paperwork, distribute industrial gases more efficiently, and secure references from satisfied customers. The driver of all of these benefits is the harnessing of electronic commerce.

Selling Business-to-Business

Electronic commerce falls into one of two categories: business-to-business sales and business-to-consumer sales. E-commerce between businesses is expected to be five times higher than business-to-consumer e-commerce. By 2003, Forrester Research Inc. estimates that business-to-business commerce will balloon to $1.3 trillion. By 2006, that figure could represent 40% of all business conducted in the United States. While both approaches share common tactics, business-to-business sales emphasize the benefits of automation and commercial relationships between buyers and sellers.

When business-to-business sales are conducted efficiently, a company gains increased time to focus on business issues, more accurate data, and quicker response times. Companies can focus on increasing revenue by extending geographic markets, improving sales channels, adding services, and growing market share. Determining current stock, shipment status of goods, and total costs becomes timely. Since information is the key to any business success, companies can manage inventory on hand, shipment cost and methods, buying patterns, distribution channels, etc. With automation, tasks can be performed almost instantly. If you deconstruct the process of ordering an item, the order form typically crosses a few desks, requires user input, and delays the transaction. With system-to-system communication, the amount of time it takes to generate an order can be reduced substantially. Combine the marketing and sales tactics of the business-to-consumer sales organizations and the automation techniques of the business-to-business sales organizations to enhance the results of your online efforts.

Ingram Micro

One of the world's largest business-to-business sales companies is Ingram Micro (www.ingrammicro.com), a wholesale distributor of technology products and services. The company distributes 225,000 products from more than 1500 hardware and software manufacturers to 140,000-plus reseller customers in 130 countries. Ingram Micro serves resellers in three market segments. In the commercial segment, the company works with independent dealers, corporate resellers, and direct marketers. In the consumer segment, the company distributes via computer and office product superstores, mass merchants, warehouse clubs, and electronics stores. In the value-added-reseller (VAR) segment, the company relies on developers of end-user solutions that add services and support to products. The Ingram Micro Web site allows resellers to research, order, and track products, request information, and receive customer service.

Marshall Industries

Marshall Industries is one of the world's largest distributors of industrial electronic components and production supplies. They provide engineering design services, complex material management systems, logistics, and information technology services to their business customers. Marshall (www.marshall.com) claims the "world's leading business commerce Web site." Marshall created one of the first industrial-strength catalogs on the Internet in 1995. Its scope is breathtaking.

Marshall sells computer products, interconnection and material handling equipment, production supplies, semiconductors, passive components, tool kits, and test equipment. Customers select among 20 separate functions, including order status, a variety of search categories, instant online help, interactive seminars, a virtual design center, sales data classified geographically, and a quoting tool. The Web site's Help@Once feature is a customer service tool that allows Marshall customer representatives to answer queries in real time throughout the day and night. In addition, customers are matched to interactive education programs sponsored by Marshall's high-tech suppliers. For almost all of Marshall's engineering customers, the Web site is the primary method of interacting with the company.

iPrint.com

Selling products and services business-to-business allows some companies to generate value for their customers as well as profits for themselves. By harnessing the power of electronic commerce, iPrint.com (www.iprint.com), an Internet-based print shop, reduces errors and costs for its clients while improving turnaround time of their print projects. iPrint.com combines a WYSIWYG (what you see is what you get) online design studio with back-office links to efficient commercial print shops, eliminating design errors and the traditional middlemen.

The online design process drives the success of iPrint.com. Customers select a product from a list that includes announcements, boxer shorts, bumper stickers, business cards, business checks, forms, envelopes, invitations, mouse pads, postcards, rubber stamps, ties, t-shirts, and lots of other choices. Then a customer selects a layout and designs the final appearance using tools incorporated into the design studio. After making additional choices about stock, ink type, and product quantity, and saving the design, the customer begins the order process. Using a Secure Sockets Layer, customers enter billing and shipping information, completing the process.

The benefits to the business customers of iPrint.com are reduced expenses, quick turnaround of projects, and reduced errors. Expenses are reduced by eliminating middlemen including intermediate print shops and design studios. Projects are completed quickly as customers no longer have to travel back and forth to suppliers and print shops. Errors are reduced because customers design and approve the composition of products without vendor intervention.

Selling Business-to-Consumer

The most common approach to electronic commerce, business-to-consumer sales, refers to the retailing of products and services to customers. Online commerce provides reasonable opportunities for brick-and-mortar companies, for online-only companies, and for the individual entrepreneur. Selling takes place directly through traditional mechanisms or indirectly through a variety of auction methods. The primary drivers for electronic commerce include reduced or eliminated brick-and-mortar costs, inventory costs, local advertising costs, and local or state sales taxes (for at least a couple of years).

Electronic commerce enhances a company's marketing focus. Marketing as a science is improved as real-time data is collected. By capturing data from your customers and creatively interpreting it, companies can improve decision making and strategic initiatives. Easy-to-navigate interfaces and creative promotions allows cross-selling and cross-promotion of products and services. Neighborhood markets become global markets with online solutions.

Electronic commerce enhances a company's product focus. Product storage and distribution costs are reduced as portions of the typical distribution chain are eliminated. Dispersed physical shelves are replaced with virtual shelves. Turnover of inventory (and the profit generated by each turn) increases as minimum stock levels decrease. Finally, overstocked items can be distributed effectively by using auction technology. The advantage here is that you control the minimum price for a product.

Electronic commerce also enhances a company's sales focus. By eliminating the need for most inventory and elaborate storefronts, costs are reduced. Some customers buy on price, some buy on convenience. As a company owner, you can target either or both. The standard virtual storefront allows a potential customer to compare brands and compare prices across your product line. With the elimination of some middlemen, savings can be passed to customers.

American Airlines

Plenty of companies that sell products and services through traditional distribution channels have created an online presence to supplement those activities. Airlines, for instance, typically initiate transactions for their service (transportation by airplane) by using travel agents or through their own agents at their branch offices. American Airlines (www.aa.com) supplements these sales and service transactions via its Web page. While the company has not ended its relationship with travel agents, online transactions do undermine the agents' gross commissions.

American Airlines' site offers a variety of attractive features. Customers can plan trips, review current reservations, and purchase tickets. They can review American Airlines' AAdvantage mileage program, AA credit card program, and business programs. Visitors can register for e-mail notice of discount tickets. Customers can also receive electronic delivery of ticket confirmations. Of secondary import, to the typical traveler anyway, is the inclusion of the inflight magazine, descriptions of the wine served onboard, and the inflight movie guide. The expense of travel agent commissions and ticket agent salaries outweighs that of American's online operations. As Internet penetration reaches further into the mainstream, anticipate the sales transactions for airlines being carried out more frequently online, instead of via traditional distribution channels.

Charles Schwab

The America's largest discount broker, Charles Schwab (www.schwab.com), offers brokerage services through its branch offices, via the telephone, and most significantly, online. Rather than adhering rigidly to its traditional branches and telephone-based business, Schwab embraced the Internet more than three years ago. The company recently reported 153,000 trades a day on its Internet site up from an average of 93,000 a day in the fourth quarter of 1998, and three times the volume of its nearest competitor. Schwab boasts over two million online accounts totaling $175 billion under management on the Web.

Schwab's comprehensive Web site offers a wide variety of components for both visitors and account holders, including planning, research, and trading services. Investors can create an investor profile, develop general financial goals, and then allocate assets to match those goals. Planning services address general, retirement, estate, college, and tax issues. In addition, visitors can request quotes, performance snapshots, charts, and professional research on most financial vehicles. Schwab allows its customers to buy and sell stocks, bonds, mutual funds, CDs, money markets, and options. Ranked #1 in online trading, breadth of products, responsiveness, and fewest extra costs, Schwab's online brokerage services demonstrate the virtue of reaching new markets with electronic commerce.

Proflowers

In order to compete in the electronic commerce arena, companies must find a niche market or employ a tactic that creates compelling results. Several companies sell flowers online with mixed results. One company, Proflowers (www.proflowers.com) provides customers with a fast, easy, and reliable way to purchase the freshest quality cut flowers at a competitive price. Proflowers competitive advantage is generated by shipping flowers directly from the grower, making bouquets and arrangements several days fresher than available in most local florists. The company guarantees the composition and quality of the bouquets with a full satisfaction money-back guarantee. Because the flowers are shipped directly from the grower, most middlemen are eliminated. Even with the shipping cost, most assortments are 30-40% below comparable retail price. Proflowers has developed a proprietary fully automated software system that provides a direct link to both FedEx and the grower. The cost savings of this system are passed on to the customer. Proflowers markets its products through an extensive network of affiliates on the Internet. The company's primary affiliate is www.bluemountain.com, the most popular electronic greeting card site. In addition, a parallel site in Spanish increases the market reach of the company. By eliminating intermediates and catering to the needs of customers, companies can venture successfully onto the Web.

Facilitating Electronic Commerce

In order to advertise online, service the customers, manage the supply chain, sell business-to-business, or sell business-to-consumer, companies rely on the products and services offered by online facilitators. These "supply and support" vendors act as intermediaries between the electronic commerce world and any companies trying to make an impact online. Facilitators provide products, such as commerce hardware and software, as well as services, such as Web site design, hosting, and marketing. The demands for these products and services continue to grow. Although several companies dominate the field including InfoSpace.com, America Online, TMP Worldwide, and MindSpring Enterprises, niche markets for online facilitators are lucrative and plentiful.

Erin Clark Design

Electronic commerce solutions can support the operations of global corporations and individual entrepreneurs. Companies that provide those solutions can also range in size from corporate giants, for example IBM, to individual consultants, for example Erin Clark Design (www.ncc.com/humans/eclark/). Erin's challenge was to leave a corporate executive position to gain control of personal commitments and projects. For a self-employed business owner like Erin, the key is to find and exploit a niche market. In her case, the niche was Internet consulting, design, and implementation. Originally the focus was strictly on designing compelling Web sites for customers. But as a result of her experience, Erin now offers complete design, development, and deployment packages. By partnering with a local ISP to provide small-to-medium sized companies with sophisticated electronic commerce solutions, Erin Clark Design helps small to medium-sized businesses move from a brick-and-mortar world to a profitable online presence. By allowing multiple clients to share platforms, the typical solution results in reduced costs to clients as hardware and software costs are spread across several companies. Erin's emphasis on continuing education and best practices allows her to effectively address business challenges and personal goals.

Vectrix

The offerings of Vectrix (www.vectrix.com), another electronic commerce solutions company, range from Web site design and development to creative services, advertising, marketing and media planning, as well as software products that enable customers to design, develop, and deploy their own e-commerce solutions. Their mission is to become a world-class e-commerce solutions company by consistently creating and delivering superior products and services for their customers. Vectrix sells its products through its Web site, a direct sales force, and distribution partners, including corporate resellers, integrators, electronic and catalog resellers, and retail outlets.

Vectrix's flagship product, EdgeworX 2000 E-Commerce Edition, is a powerful, affordable, easy-to-use software tool that helps businesses sell products and services on the Internet. EdgeworX incorporates e-commerce design and component features, advanced development features, administration features, and open deployment and access features. The software application includes a wizard that guides users through the process of creating catalogs, product listings, shopping carts, shipping and handling fee calculations, checkout processing, and order processing. EdgeworX 2000 E-Commerce Edition offers open system architecture and support for the latest Web technology standards, including database compatibility, Web design, server platform independence, browser flexibility, and system security. In addition, Vectrix enhances the value of this e-commerce product by providing exceptional customer service.

ShopInCities International

Another category of online facilitators is the product/service aggregator. An aggregator offers companies a storefront in a virtual mall. While traditional malls rely on the health and prosperity of their anchor tenant, virtual malls are limited only by bandwidth and the quality of the interface. The client companies of aggregators can focus on business-to-business sales or on business-to-consumer sales. ShopInCities International, a dominant product and service aggregator, allows client companies to create fully functional e-commerce storefronts. After these storefronts are designed and developed (using software tools available on the aggregator's Web site), they are placed in their various home cities allowing customers to shop for products from their own hometown or from their favorite global destinations. ShopInCities International plans to deploy over 300 city sites by the end of 1999.

The advantage to clients of an aggregator are plentiful. By refocusing the Internet shopping experience from a large conglomeration of individual business sites to community shopping malls of e-commerce storefronts, aggregators allow potential customers to more conveniently search for products and services. Client companies are able through the mall concept to pool the resources of fellow storefronts and the expertise of the aggregator. More stores mean more eyeballs. More eyeballs mean more sales. The real benefit generated by an aggregator is that their mall makes small companies big and makes big companies small.

Some Words of Caution

Companies have enjoyed success when advertising, servicing customers, managing their supply chains, selling business-to-business, selling business-to-consumer, and also, developing the infrastructure of other companies. For all of the advantages that electronic commerce offers, some disadvantages still trouble the medium. From a merchant's perspective, your customers still want to touch and feel your product. They still seek the instant gratification of walking out of the store with a bag full of your product. They still distrust the electronic interface, the miles that separate your store from their computer. From a business perspective, companies should expect fairly steep startup expenses, maintenance costs, and training costs. Employees will be laid off or reassigned as labor and skill requirements change. When you make mistakes with your online efforts you risk wallet and reputation. eBay recently suffered a one day long outage of their auction service resulting in a $4 billion reduction in their market capitalization. The constant in electronic commerce is global competition. Other companies work hard to steal away your customers. These competitors are not going away. Strike back at them.

Most companies capitalize on only a couple of these electronic commerce approaches. If they ignore the online sales channel because of established distributor relationships or a lack of expertise, they may still generate positive results by advertising online. Companies that focus their energies advertising online may eventually choose to add a customer service component. Those companies that sell to businesses or consumers may enhance their performance by automating their supply chains. Challenge the third-party providers and the facilitators to continue to deliver innovative products and services. Revise your business model to incorporate online advertising, customer service, supply chain management, and sales. As you absorb the electronic commerce approaches and the representative companies, turn your attention to the architecture that generates an effective electronic commerce solution.

5

Architecture

- What components comprise an architecture?
- Performance tuning and load balancing
- Network architectures
- Architecture security
- Architectures in the real world

Even with the emphasis placed on the business focus of e-commerce solution development, an understanding of the underlying technologies is critical. Few individuals will actually have responsibility for implementing or installing the software required to deploy a solution. However, knowledge of product functionality, limitations, and technical requirements offers a measurable competitive advantage.

Overview of Components

The components that comprise the typical electronic commerce solution are Web clients, Web server software, commerce server software, connectivity tools, and back-end systems. Learn the features and functions of each of these components, as well as how they relate to each other. In later chapters, we address vendor solutions, payment systems, and security considerations, each focused on the component nature of a solution. The vocabulary associated with these five components enhances your understanding of the design process and proves invaluable as you select software and hardware packages for implementation. Figure 5.1 illustrates a typical e-commerce architecture.

Figure 5.1 A Typical E-Commerce Architecture

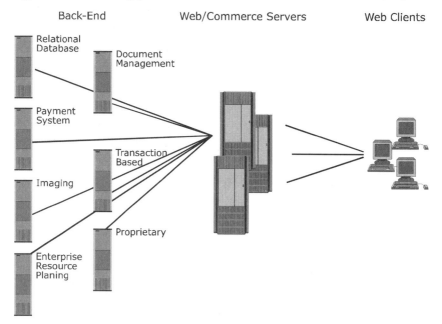

Back-End   Web/Commerce Servers   Web Clients

Relational Database
Document Management
Payment System
Transaction Based
Imaging
Proprietary
Enterprise Resource Planing

Web server software publishes hypermedia documents on-the-fly for the Web client interface. Commerce server software generates and supports storefront functionality for merchants. The Web server software and the commerce server software reside on the same hardware component. Customers access the vendor site via Web clients (or as they are more commonly known, Web browsers). Back-end systems, which typically already exist in an organization, require a connectivity tool to link them to the server software. This linking allows companies to leverage data such as product information, customer information, and transaction records.

E-commerce is historically a direct descendent of electronic data interchange (EDI). This technology is used to communicate between large companies. However, standards, based on the specific industry, are required for the exchange of information (data) electronically between computer systems. For example, the automotive industry requires that each participant use the Automotive Industry Action Group (AIAG) standard to communicate. Grocery retailers are required to use the Uniform Communications Standard (UCS). Furthermore, all participants are required to be connected to the same private network. So, how do the smaller companies participate? This is one of the major problems. EDI is very costly to implement and difficult to provide interoperability between different industries, since each industry has its own standard for data exchange.

For those companies that currently communicate via EDI, e-commerce software vendors provide solutions which allow the integration of existing EDI systems. Microsoft, IBM, Lotus, Netscape, and other software vendors provide EDI extensions as part of their tool sets to integrate into existing EDI-based systems. As part of these solutions, companies that were restricted to automating processes between those business partners that follow the EDI standards can now send information such as purchase orders directly to smaller business partners via a Web browser.

The electronic commerce architecture, in comparison, allows companies to communicate using one standard communication protocol, called TCP/IP (Transmission Control Protocol/Internet Protocol) and a data format called HTML (Hyper Text Markup Language). Think of a protocol as a language. In order for two people to speak and then understand each other, they need to communicate in the same language. What happens when one person speaks and understands only German and the other speaks and understands only Italian? They need a translator. The communication between these individuals requires some additional time and effort. In addition, the transaction costs a little more since the translator must be paid. That is the beauty of communicating in one common language...no additional translation cost. Electronic commerce has enabled interoperability between different business systems, increasing efficiency and lowering costs. Businesses can now easily communicate with each other through Web servers, and with consumers through Web browsers. That is why we are seeing a big shift in the way business is being conducted on a city, state, country, and worldwide basis. Everyone speaks the same language.

Web Clients

A Web client (from the perspective of the merchant) or browser (from the perspective of the customer) is an application program that provides a graphical interface to view and interact with all the information available on the World Wide Web. The word "browser" originated as a generic term for user interfaces that let you browse text files online. As Web content began to also include graphics, the term became more common. Technically, a Web browser is a client program that uses the Hypertext Transfer Protocol (HTTP) to make requests of Web servers throughout the Internet on behalf of the browser user. A commercial version of the original browser, Mosaic, is still popular today. This version, including many of the user interface features in Mosaic, is Netscape Navigator. As Microsoft began to realize the growing popularity of the Web, it developed and freely distributed Internet Explorer. Although the online services, such as America Online, Compuserve, and Prodigy, originally featured their own browsers, virtually all providers now offer the Netscape or Microsoft browser. Other common browsers include Lynx, a text-only browser for UNIX shell and VMS users, and Opera, a sophisticated, yet compact Web browser.

An important component of the development process for an electronic commerce solution is testing your resulting Web site pages, embedded code, and scripts on the dominant browsers available to the public. Currently, those browsers are Microsoft's Internet Explorer and Netscape's Navigator (or the Communicator suite). Microsoft commands a 73% market share to Netscape's 25% with 2% shared among several others. Statistics on browser usage are available, and updated daily, at www.statmarket.com. Always build solutions for the latest proven version of the defacto standard browser.

Web Server Software

Web server software serves as a middleman between back-end systems and front-end Web clients. Its primary function is to generate and deliver hypermedia documents based on Hypertext Markup Language (HTML). In most cases, the HTML code is generated on-the-fly by the Web server software. The World Wide Web Consortium (www.w3.org) sanctions the latest standards of HTML. Vendors of these Web server products support a variety of network operating systems, including NT, UNIX, Novell, and OS/2. Web server candidates for your electronic commerce solution include Microsoft Internet Information Server (IIS), Domino Server, Netscape, and Apache Server. Figure 5.2 illustrates the Web server/commerce server components and their relationship to Web client software. For more information about Web server software, its functionality and suitability, visit the major vendors' Web sites.

Figure 5.2 Web Server/Commerce Server

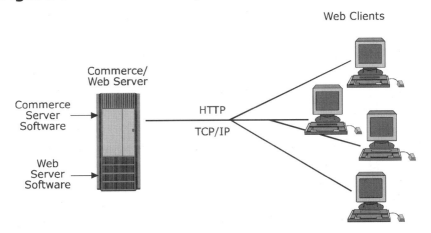

Commerce Server Software

At the heart of an e-commerce solution lies the commerce server software, with capabilities for creating an online storefront. Several major vendors on the market offer a variety of software solutions that allows you to establish an e-commerce site on the Web. The components of this software include storefront implementation tools, commerce server management tools, and back-end integration tools.

Many commerce server software solutions include standard store templates that can be readily customized or they include wizards to walk you through the construction phases of building a store. Several vendors have sample stores that you can use as a basis for the site. The selected template serves as a model of your eventual store and represents the hub of your e-commerce site. Vendor-oriented storefront features include product administration, inventory tracking, purchase order generation, credit card verification, sales tax and shipping calculation, and site analysis tools. Customer-oriented storefront features include product listings, discount pricing/auction technology, question and answer listings, product comparisons, order tracking, shipment tracking, and search tools.

Advertising features include registration, cross-selling, promotion tools, push technologies, and mail distribution. Registering visitors is extremely important to your company because it allows you to capture data about your customer. This data can later be used to analyze buying trends, send product information back to your customer, and set up personalized options based on their preferences. Cross-selling features use shopping trends and existing transaction information to recommend products of interest to customers. Advertising capabilities may include banners, featured products, and special discounts. Push technology provides users with channels that deliver specific content. The user subscribes to a channel and related information is pushed (delivered) to the browser on the desktop. Now commerce vendors provide solutions featuring push technology as one of their delivery mechanisms. Direct mail is similar to an electronic version of paper-based direct mail. The difference is that you can use direct mail to deliver personalized content to thousands of users with lower costs than paper-based delivery. In addition, direct mail offers automated delivery, a standard for Internet marketing.

Most vendors include the following server management capabilities: content management, replication and clustering, site usage statistics, and remote administration. Content management is a feature which allows you to manage your content, including new products, the design of pages, and additional categories. Replication and clustering capabilities are standard features of the commerce software or management component. In order to determine when to scale your site, employ site usage tools to provide you with a mechanism to track server performance. Commerce programs provide the tools needed to ensure that your servers perform efficiently.

Linking your existing inventory management and finance systems to your commerce software remains one of the most challenging aspects of setting up a successful e-commerce site. Before selecting a commerce server software package, make sure that you can easily integrate it with your existing back-end systems.

Connectivity Tools

Connectivity tools act as translators to connect back-end systems to server software or front-end clients (browsers). Early e-commerce Web sites required the use of programming languages, such as C++, to perform the appropriate data translations that link these components. Today, many commerce software providers and third-party vendors provide tools that offer connectivity with minimal programming efforts. These tools are designed typically to push and pull data from the Web/Commerce server to the back-end system.

Back-End Systems

When designing and deploying an e-commerce solution, make sure that you can easily integrate your existing back-end systems. Determine exactly what systems you have in place. Your commerce solution requires that you pull data from these systems to your commerce site. Gather the names of the software vendor and the software packages (with version numbers). In addition, you need to understand your current infrastructure. Back-end systems may include relational databases, transaction-based systems, ERP systems, EDI, third-party software, and proprietary systems.

Figure 5.3 Back-end Systems

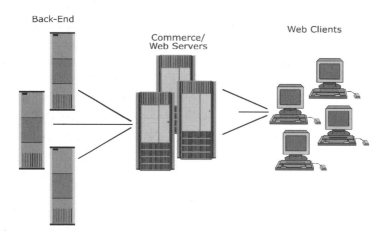

Integration with existing relational databases is provided in most commerce package solutions. Each vendor has its own set of tools that provides integration to specific relational databases. The question you must answer is how easily a commerce system can be integrated. Commonly supported databases include Microsoft's SQL, IBM's DB2, and Oracle's relational databases. Even if there is not an "out of the box" solution, some other database vendors, such as Sybase, have built tool sets that you can use to tie into these systems. Some database vendors have code at their site that you can use as a starting point to tie into these system. Begin your research at the commerce system vendor's site to determine if they include a tool to support integration.

The goals of the integration of commerce systems to transaction-based systems are seamless processing of orders and timely, accurate updates. Most commerce vendors offer connections to a variety of transaction-based systems, although with varying degrees of compatibility. The questions you must address in your research include how difficult is the implementation and how robust is the solution.

The focus of IBM and Lotus includes integration into any back-end environment, especially legacy platforms (CICS and IMS) and transactional middleware (MQSeries). Their solutions have grown more robust and user-friendly with each release. Integration tools include database tables, scripts, sample code, and documentation to support transactions and message queues with other environments. In contrast, some vendors, including Microsoft, have not focused on providing the same breadth of connectivity.

Most vendors provide solutions to integrate with some of the most common enterprise resource planning (ERP) systems, including PeopleSoft and SAP. Commerce server systems add value to your existing ERP system by passing only fully formatted orders, by acting as a firewall, by eliminating direct access of Internet users, and by enhancing the catalogs presented to customers with more interactivity.

If you have an accounting package, tax software or any other type of system that you want to integrate into the commerce environment, go to the commerce vendors and the third-party vendors to determine if an integration tool is available. If it is not, you must write the API to perform the integration.

Electronic data interchange (EDI) allows program-to-program data exchange over private or public networks. It is based on a set of predefined relationships and standards between the participating parties. These predefined relationships and standards include interfaces, translators, consistent data, and mapping protocols. Commerce server solutions support electronic data interchange by delivering merchant orders or customer special orders to the supplier, and allowing suppliers to report order status to the merchant.

Connecting a commerce system to a proprietary software system requires building an application programming interface (API). An API serves as a middleman by making requests between the two systems. For example, many companies use a proprietary credit card verification system. An API connects the commerce server to that proprietary system, ensuring efficient and accurate communication.

Tuning and Load Balancing

Your Web site might perform adequately as soon as it is deployed. It might continue to function effectively for days or even weeks. At some point, the demands made on the system will overtax it, reducing its performance. Servers deliver Web pages with increasing delay. Some pages are not even available. Site traffic bogs down the system. And then the traffic vanishes. Customers, frustrated with the delays and navigational obstacles, stop visiting your site.

The performance of your electronic commerce solution is typically measured by its fault tolerance and load balance. Fault tolerance refers to the capacity of a site to continue to provide service, even when a server node fails. A fault tolerant configuration features server nodes that immediately pick up the requested load with minimal disruption to visitors when a server node stops working. The process of automatically transferring workload from one server node to another server node is called failover. The process of restoring the workload back to the original server node is called failback. Failover and failback are processes that clustering applications perform automatically and transparently, so that the visitor is typically unaware of their occurrence.

Load balance indicates that the amount of traffic on an individual server within a group of servers is relatively equal to the demand placed on each of the group members. Rather than overloading one server node and underloading another server node, the load can be handled efficiently by distributing the load between the two. Load balance can be achieved by assigning Web sites to a specific server node manually. This configuration allows a Web site to be accessed from either node, although a specific Web site may not be accessed from both nodes simultaneously.

Clustering supports load balance of a system, as well as fault tolerance capabilities. Replication supports quicker retrieval of data by users and fault tolerant operation of the e-commerce solution. Software vendors incorporate both of these features into most commerce packages. An additional component (either software or hardware), called a router, helps balance the load demand for individual servers. The following section describes each of these enhancements to your system.

Clustering

Clustering allows two servers, with real-time updating, to appear as one to site visitors. The servers are connected not only physically by cables, but also programmatically, with clustering software. This connection allows the servers to support failover and failback functionality, as well as load balancing. Clustering software is typically included with the commerce package.

Figure 5.4 Clustered Servers

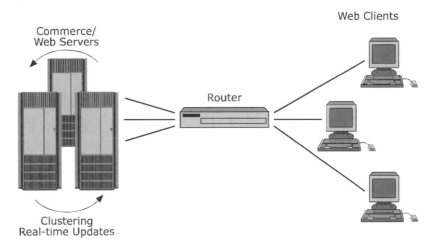

Replication

Replication refers to the copying of configuration information and content from one server to another so that both servers offer the same resources to visitors. The site administrator determines and schedules the frequency of further updates of data. The advantages of replication are improved performance and better availability. Applications can operate on local copies instead of having to communicate with remote sites, improving performance. A replicated server remains available for processing as long as at least one copy remains available. The primary disadvantage is that when a replicated server is updated, all copies must be updated. Replication is supported by most commerce packages.

Figure 5.5 Replicated Servers

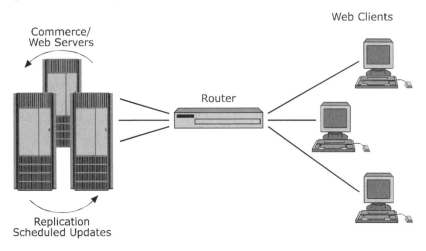

Routers

In both of the previous diagrams, a router sits between Web/ Commerce servers and Web clients. Although these diagrams are simplified (packets of information travel to and from other servers across the Internet before reaching clients), the position of the router is accurate. A router is a hardware device or, in some cases, software incorporated into the server, that determines the next network point to which a packet of data should be forwarded toward its destination. The router is connected to at least two network pathways and determines which pathway to route each information packet based on its understanding of the state of the networks.

A router creates or maintains a table of available routes and their conditions. The router then combines this information with distance and cost algorithms to determine the best route for a specific packet. From the perspective of the commerce package, the advantage of routers is their load balancing functionality. Leading manufacturers of routers include Ascend Communications, Cisco, and 3Com. More information about routers and their implementation is available at the vendors' sites.

Network Architecture

While the concept of architecture within this chapter primarily refers to the functionality and relationship of electronic commerce components, a higher level perspective concerns the actual configuration of a network. Any network can be classified as one of three arrangements: Internet, intranet, or extranet. The significance of this classification is that it establishes the audience, security considerations, and performance levels that your commerce solution must address.

Internet

The Internet, usually called the "Net," is a worldwide system of computer networks, a network of networks in which users at any one computer can, with permission, get information from any other computer. It was conceived by the Advanced Research Projects Agency (ARPA) of the U.S. government in 1969 and was first known as the ARPANet. The original aim was to create a network that would allow users of a research computer at one university to be able to communicate with research computers at other universities. A side benefit of ARPANet's design was that, because messages could be routed or rerouted in more than one direction, the network could continue to function even if parts of it were destroyed.

Today, the Internet is a public, cooperative, and self-sustaining facility accessible to hundreds of millions of people worldwide. Typical users include company employees, government workers, researchers at educational institutions, and private individuals. Physically, the Internet uses a portion of the total resources of the currently existing public telecommunication networks. Technically, what distinguishes the Internet is its use of a set of protocols called TCP/IP (Transmission Control Protocol/Internet Protocol).

Two recent adaptations of Internet technology, the intranet and the extranet, also use the TCP/IP protocol.

The most widely used part of the Internet is the World Wide Web (often abbreviated "WWW" or called the "Web"). Its compelling feature is hypertext, a method of instant cross-referencing. In most Web sites, certain words or phrases appear in text of a different color than the rest, and underlined. When users select one of these words or phrases, they are transferred to the site or page that is associated with this word or phrase. Sometimes there are buttons, images, or portions of images that are "clickable." If you move the pointer over a spot on a Web site, the pointer changes into a hand, indicating that you can click and be transferred to another site.

The Web allows access to millions of pages of information. Web surfing (moving from page to linked page) requires a Web browser, the most popular of which are Netscape Navigator and Microsoft Internet Explorer. The appearance of a particular Web site may vary slightly depending on the browser used. Also, later versions of a particular browser are able to render more bells and whistles such as animation, virtual reality, sound, and music files, than earlier versions.

Intranet

An intranet is a series of networks contained within an enterprise. It may consist of many interlinked local area networks (LAN) and also relies on leased lines in the wide area network (WAN). Typically, an intranet includes connections through one or more gateway computers to the Internet. The main purpose of an intranet is to share company information and computing resources among employees. An intranet can also facilitate working in groups and teleconferences. Web browsers allow companies to deploy the intranet infrastructure by providing a graphical user interface (GUI) that is inexpensive and easy to deploy.

An intranet uses TCP/IP, HTTP, and other Internet protocols and in general resembles a private version of the Internet. With tunneling, companies can send private messages through the public network with special encryption/ decryption and other security safeguards to connect one part of their intranet to another part.

Typically, larger enterprises allow users within their intranet to access the public Internet through firewall servers that have the ability to screen messages in both directions to maintain company security. When part of an intranet is made accessible to customers, partners, suppliers, or others outside the company, that part is an extranet.

Extranet

An extranet is a private network that uses Internet protocols and the public telecommunication system to securely share part of a business' information or operations with suppliers, vendors, partners, customers, or other businesses. Think of an extranet as part of a company's intranet that is extended to users outside the company. The same benefits that HTML, HTTP, SMTP, and other Internet technologies have brought to the Internet and to corporate intranets now facilitate business-to-business transactions.

An extranet requires security and privacy. These characteristics require firewall server management, the issuance and use of digital certificates or similar means of user authentication, encryption of messages, and the use of virtual private networks (VPNs) that tunnel through the public network.

Companies can use an extranet to exchange data using electronic data interchange (EDI), allow access to product catalogs exclusively for wholesalers, collaborate with other companies on joint development efforts, and develop and use training programs with other companies. They can also provide or access services provided by one company to a group of other companies, such as an online banking application managed by one company on behalf of affiliated banks, and share news of common interest exclusively with partner companies.

There are a variety of configuration options for your commerce solution. Each has its own advantages. Figure 5.6

illustrates a typical configuration, with back-end systems, Web/Commerce servers, routers, and front-end clients.

Figure 5.6 A Typical Configuration

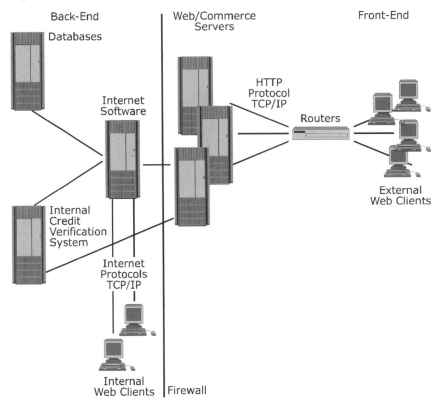

In Figure 5.7, replication is used to synchronize the databases from the external Web/Commerce server to the internal Web/Commerce server. This configuration provides an extra layer of security by allowing the site administrator to select which Web pages to make available to other companies and individuals. The other advantages of this configuration are easier management of all server-to-server replication, easier Web application management (only select applications are posted externally), and enhanced security, as internal and external audiences are separated.

Figure 5.7 Another Typical Configuration

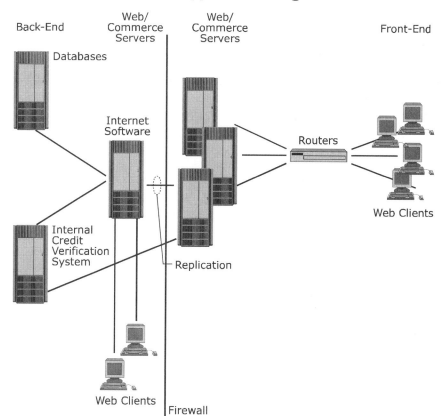

Architecture Security

Protecting your system components typically requires both firewalls and proxy servers. Firewalls protect the resources of a private network from users of other networks. Proxy servers ensure security, administrative control, and caching service. A proxy server is associated with or part of a gateway server that separates the enterprise network from the outside network and a firewall server that protects the enterprise network from outside intrusion.

Firewalls

A firewall is a set of related programs, located at a network gateway server, that protects the resources of a private network from users of other networks. The term also implies the security policy that is used with the programs. An enterprise with an intranet that allows its workers access to the Internet installs a firewall to prevent outsiders from accessing its own private data resources and for controlling what outside resources its own users can access.

A firewall, in conjunction with a router program, filters all network packets to determine whether to forward them toward their destination. A firewall also includes or works with a proxy server that makes network requests on behalf of workstation users. A firewall is often installed in a specially designated computer separate from the rest of the network so that no incoming request can directly access private network resources.

There are a number of firewall screening methods. A simple one is to screen requests to make sure they come from acceptable, previously identified IP addresses and domain names. For mobile users, firewalls allow remote access into the private network with secure logon procedures and authentication certificates.

A number of companies make firewall products. Features include logging and reporting, automatic alarms at given levels of intrusion, and a graphical user interface for administrators.

Proxy Servers

For an enterprise that accesses the Internet, a proxy server is a server that acts as an intermediary between a workstation user and the Internet so that the enterprise can ensure security, administrative control, and caching service. A proxy server separates the enterprise network from the outside network, while a firewall server protects the enterprise network from outside intrusion.

A proxy server receives a request for an Internet service, such as a Web page request, from a user. If it passes filtering requirements, the proxy server, if it is also a cache server, looks in its local cache of previously downloaded Web pages. If it locates the page, it returns it to the user without needing to forward the request to the Internet. If the page is not cached, the proxy server, acting as a client for the user, uses one of its own IP addresses to request the page from the server out on the Internet. When the page returns, the proxy server matches it to the original request and forwards it to the user. Proxies mask the return address of the requesting computer, providing anonymity for users.

An advantage of a proxy server is that its cache can serve all users. If one or more Internet sites are frequently requested, these are likely cached by the proxy, which reduces user response time. A proxy can also log transactions between it and its user group. In addition, some proxies run virus detection programs on incoming packets.

The functions of proxy, firewall, and caching can be in separate server programs or incorporated into a single package. Different server programs can be in different computers. For example, a proxy server may be located in the same machine with a firewall server or it may be on a separate server and forward requests through the firewall.

E-Commerce Architecture Models

Let's take a look at some companies that have designed, developed, and deployed electronic commerce solutions. The profiles list the components that each company employs, describe unique characteristics of the implementation, and illustrate how the components are related. Reviewing their architectures helps you better understand how each component fits into the system. Use these profiles as models for your own solution.

CERA Bank

CERA Bank is an independent full-service bank with a network of 950 branches and more than 4,600 staff members, making it the sixth largest financial institution in

Belgium. CERA Bank's hardware and software components (see Figure 5.8) include firewalls, IBM Internet Connection Server, MQSeries, MS Database, customized encryption, Sun Microsystems Java applet calculator, and OS/2 Warp Server.

MQSeries receives the incoming data from the IBM Internet Connection Server and sends it to the MS database. MQSeries then batches the responses and sends them back through the IBM Internet Connection Server. MS Database holds the logic, applications, and data.

IBM Internet Connection Server hosts the HTML pages and Java applet. Customized encryption software is used between the internal systems and the client. OS/2 Warp Server hosts the IBM Internet Connection Server. The Java calculator application allows the customer to perform mortgage calculations on the Web client.

Front-end components include all popular Web clients.

Figure 5.8 CERA Bank Architecture

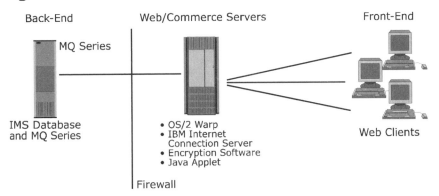

Back-End   Web/Commerce Servers   Front-End

MQ Series

IMS Database   • OS/2 Warp
and MQ Series   • IBM Internet   Web Clients
                        Connection Server
                      • Encryption Software
                      • Java Applet

Firewall

Dell Computers

Dell is the world's largest direct PC company with more than 26,100 employees in 33 countries. Dell's hardware and software components (see Figure 5.9) include Microsoft Site Server Commerce Edition, Microsoft Windows NT Server with Internet Information Server, Microsoft SQL Server, Microsoft Visual SourceSafe, Dell PowerEdge Servers with dual 200 MHZ Pentium Pro processors, Tandem Servers, internal credit card verification system, NetWeave, and Cisco Router and Distributed Director.

Microsoft SQL Servers are used to retrieve legacy data from a Tandem mainframe using NetWeave. Microsoft SQL Servers are also used to track the progress of orders and Premier Pages which contains specific client information, such as order history, account contacts, and system configurations for clients. The credit verification and authorization system is internal to Dell.

Internet Information Server (IIS) provides the engine that serves information to the customer. IIS is a Web server platform that includes Internet services such as Web access and FTP. Microsoft Site Server Commerce Edition is used to conduct transactions from the customers to the business. In addition, the software replicates the Web content to the other servers twice a day. This allows distribution of load between the Web servers.

Figure 5.9 Dell Architecture

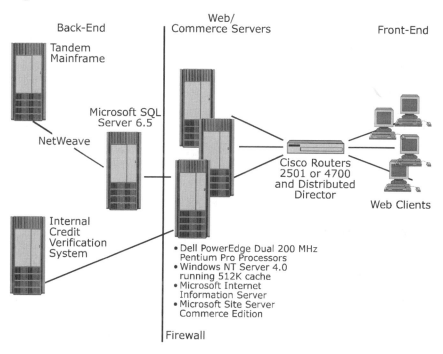

Back-End

Tandem Mainframe

Microsoft SQL Server 6.5

NetWeave

Internal Credit Verification System

Web/ Commerce Servers

- Dell PowerEdge Dual 200 MHz Pentium Pro Processors
- Windows NT Server 4.0 running 512K cache
- Microsoft Internet Information Server
- Microsoft Site Server Commerce Edition

Firewall

Cisco Routers 2501 or 4700 and Distributed Director

Front-End

Web Clients

Microsoft Internet Explorer push technology is used to send customers updates and notices about new products. The Cisco Router and Distributed Director are used to balance the load across the front-end Web servers.

Dreyfus Brokerage Services, Inc.

Dreyfus Brokerage Services, Inc. is a popular brokerage company that allows thousands of investors to trade at their convenience via the Web. An average of 600,000 to 700,000 investors hit Dreyfus Brokerage's Web site each day. Dreyfus Brokerage's hardware and software components (see Figure 5.10) include IBM DB2, IBM OS/400, IBM DB2/Connect, AS/400 as the Web server, I/NET Commerce Server/400, and internal databases.

IBM DB2 databases contain customer profiles, name and address data, real money balances, security holdings, and quote files. I/Net Commerce Server/400 is the Web server. IBM DB/2 Connect is used to update records between I/Net Commerce and the IBM DB/2 back-end databases. Front-end components include all popular Web clients.

Figure 5.10 Dreyfus Brokerage Architecture

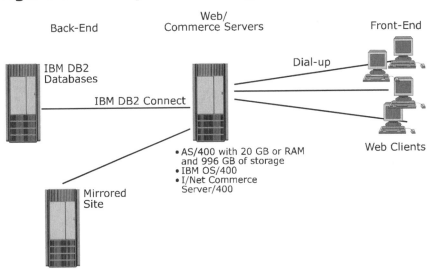

Holt Educational Outlet

Holt Educational Outlet is a retailer of high quality educational toys featuring 20,000 products online with search engine capabilities. The system is a Web-based electronic-commerce system with integration into a back-end financial system. Holt's hardware and software components (see Figure 5.11) include Pentium Pro ALR Server, Microsoft Site Server Commerce Edition, Microsoft Windows NT Server with the Internet Information Server, Microsoft Transaction Server, Microsoft SQL Server, Microsoft Proxy Server, Great Plains Dynamics C/S+ Financial Modules for its Tax System, VeriFone vPOS for credit card processing, and Cisco 2500 Routers.

Microsoft SQL Server is the backbone for the Web site and its financial systems. A primary SQL server contains 10 GB of information and a data warehouse with another 20 GB of data. It is used to hold the product information and prices. VeriFone vPOS credit verification/authorization software is used for credit card verification using SET 1.0. Great Plains Dynamics C/S + is used as the tax application. The financial modules contain general ledgers, receivables management, and inventory control. Dynamic Commerce is an application that links Microsoft Site Server, Commerce Server, and SQL Server to Dynamics C/S+.

Internet Information Server (IIS) provides the engine that serves information to the customer. IIS is a Web server platform that includes Internet services such as Web access and FTP. IIS is included in Windows NT Servers. Microsoft Site Server Commerce Edition is used to handle the transactions and to integrate the Web-based orders directly into Great Plains Dynamic Commerce and Dynamic C/S+ financial management software. Integration between the systems is performed in real time. This allows customers to view how many items are in stock. The analysis feature is used to produce reports such as Web sales and number of hits per site. Microsoft's Transaction Server is part of the NT Server Option Pack. The software ensures integrity of transactions that involve multiple systems. For example, the software would ensure that customers are not billed for merchandise that is not in stock.

Front-end components include all popular Web clients. Microsoft's Proxy Server functions as the firewall between the Intranet and Extranet. The Cisco Router is used to connect the internal and external networks.

Figure 5.11 Holt Architecture

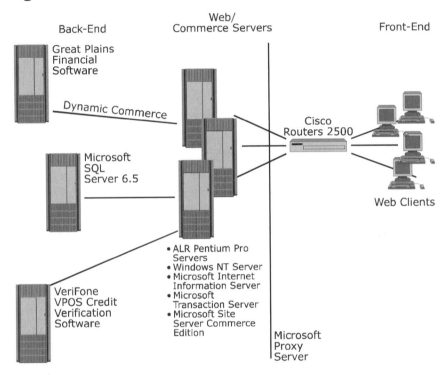

QCS

QCS is an electronic community that helps retail buyers collaborate with their worldwide supply chain. QCS hardware and software components (see Figure 5.12) include Domino Server, Domino Applications, Web clients, and Notes clients.

Each Supplier has its own back-end system and electronic payment system. Web/Commerce server components include Domino servers. Front-end components include all popular Web clients.

Figure 5.12 QCS Architecture

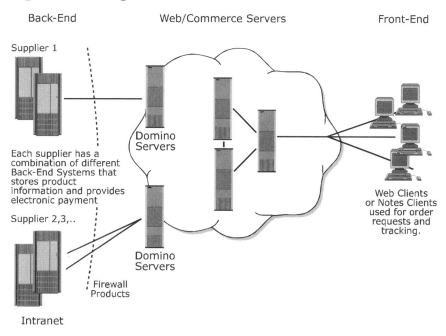

The Shell Group

The Royal Dutch/Shell Group of Companies, usually known as Shell, has grown out of an alliance between Royal Dutch Petroleum Company in the Netherlands and the Shell Transport and Trading Company in the UK. With a Supplier Managed Inventory System (SMI), Shell manages inventory for some of its customers. In return, the customers agreed to commit to Shell as their only supplier. Shell's hardware and software components (see Figure 5.13) include Domino Server, Domino Applications, Lotus Notes, Web client, SAP, and SAP Connectivity Tool.

SAP loads all customer data from the Notes server to the SAP system as a batch job. SAP then calculates a new supply plan for the client. Within SAP the Purchase Order is created. Domino Server converts the Notes data into HTML on-the-fly for Web clients. Domino Application is an inventory system used to manage customer orders. The Notes client is placed at the customer site and replicates with the Shell Notes server to determine daily consumption, current inventory levels, status of products arriving, and customer forecasts. Customers that do not have Notes use the Web client.

Figure 5.13 Shell Architecture

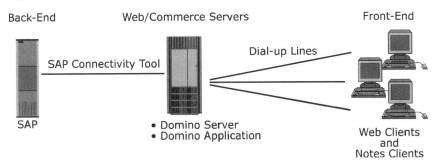

With an understanding of electronic commerce components, performance enhancements, network configurations, and security software, we turn our attention to the heart of any compelling solution, the commerce server software.

6

Commerce Server

- Implementing storefronts
- Storefront features
- Commerce server management tools
- Integrating back-end systems
- Customizing commerce server software
- Skills needed to configure commerce server software

S everal vendors on the market offer a variety of software solutions that allow you to establish an electronic commerce site on the Web. In order to select among these solutions, you must understand how your company intends to participate in the e-commerce marketplace. Evaluate your buyers and suppliers, and formulate new marketing strategies. Although most vendors offer standard options for creating storefronts, based on your business needs, some may fit your business processes much better than other vendor solutions. Questions to ask include:

- Are your going to sell products or services or deliver digital content?
- How are you going to advertise products?
- Are you going to use auction technologies?
- Are you going to deploy a storefront or an e-commerce solution that follows a completely different business model?

Second, you must understand your company's technology infrastructure and IT skill sets. The goal is to get your e-commerce site up and running quickly, efficiently, and with the least amount of expense. When selecting an e-commerce software package, it is important to review the following:

- What back-end systems do you need to integrate into the new commerce site? This may include relational data-bases that store existing product data, transaction-based systems, enterprise resource planning systems, and accounting software.

- What skill sets are available within my company to deploy this solution?

- Should I outsource or use internal personnel?

This chapter provides you with a foundation to address these areas with an understanding of your options. This knowledge proves invaluable when comparing commerce solutions. Specifically focus on these main components: storefront implementation, storefront features, commerce server management tools, integration into existing systems, flexibility to customize, and skill sets required to deploy the solution. In this chapter we review some of the major ven-dors of commerce software to see what features they support with their products.

Storefront Implementation

The storefront is the focal point for selling your products and services. Storefronts may incorporate and support prod-ucts for sale, services provided, company information, job postings, special events, featured products, store registra-tion, and other elements that add value to your customer's experience at your site.

Selling Products

The most common shopping model on the Web involves using storefronts to sell products. When buying products, the customer selects the item to purchase and adds it to a shopping cart. Once the customer has finished shopping, the customer pays for the item online using a credit card. After the credit card is approved, the system sends the customer a confirmation and the items are delivered. The concept is very similar to the product catalogs currently used by thousands of companies. Each product has a product SKU, name, description, and price. Other features may include color, size, weight, images of the product, or any other attribute that is specific to the product you are selling.

Selling Digital Content

As available bandwidth to the consumer increases, electronic delivery of digital content becomes more practical and more popular. Most commerce software does not provide an "out-of the box" solution to support the sale and delivery of electronic goods that customers download. However, the major solutions do support the purchase side of the transaction. Typically, a third-party component is required to support the delivery side of the transaction. This chapter addresses only the purchase side of the transaction, a common ground for the major commerce software components.

Storefront Features

Many e-commerce software solutions include standard store templates that can be readily customized or wizards to walk you through the construction phases of building a store. Several vendors have sample stores that you can use as a basis for the site. The selected template serves as a model of your eventual store and represents the hub of your e-commerce site.

After you walk through the wizard screens, the site still requires modification. The most common question about these wizards is whether people really use them. The answer is yes, if they are creating a storefront to sell products or services. The rationale is that these wizards contain the basic information you need to create the site, such as the

storefront home page, product information, tax and shipping information, credit card information, design, and layout. These wizards also allow you to easily integrate third-
party credit card verification systems (such as Cybercash or
VeriFone), Taxware International services, UPS tracking
capabilities, and inventory tracking systems. The integration of these third-party components represents the most
compelling reason to at least consider the wizards.

Why reinvent the wheel? It is much easier to create a
store using these tools than it is to construct a store from
"scratch." If you build your solution from the ground floor
up, it may take several developers several months to reach
the same outcome as the wizard-based solution offers. Put
your efforts where your expertise is most developed. Where
it is less developed, harness the expertise of others. In the
field of e-commerce, time is your primary opponent. Do not
spend time learning a skill that you can rent instead.

When selecting a commerce solution, it is important to
review the built-in templates or wizards which are part of
the product. Each vendor has solutions that vary slightly in
their out of the box capabilities. By purchasing a package
that meets your business needs, you can reduce the amount
of time and resources required to customize a solution. You
must compare the cost of a standard software solution with
that of a custom solution designed and developed by internal or external resources.

Vendor-Oriented Features

The following section addresses commerce software features
from the perspective of the vendor. These features include
product administration, inventory tracking, purchase order
generation, credit card verification, sales tax and shipping
calculation, and site analysis tools.

After you have designed, developed, and deployed your
store, your focus falls on administration of the store's products. The easier, the better. Several e-commerce solutions
allow you to tweak the commerce component of your site via
any common Web browser. Administration includes capabilities such as changing product properties with point-and-

click expedience, as well as adding, updating, and deleting products.

The ability to track inventory varies from solution to solution. Some vendors offer the capability to integrate your commerce site into existing inventory tracking systems. However, most vendor solutions stop short of supporting order tracking. What happens to stock on hand after an order? To take the next logical step, most systems require customization to tie into existing inventory tracking systems. As support for this integration seems to be a weak point with most of this commerce software, you are likely to be forced to build your own solution.

Your solution, since it is developed from the ground floor up, may track incoming products, items ordered by customers, and stock on hand. For new companies, you need to create your traditional and your online transaction systems. This means creating additional tables to store data, integrating the items ordered from the online site, building an inventory system to track items ordered, assessing stock on hand, determining low supply of stock, and establishing reorder levels.

Purchase order (PO) generation is a standard capability for most commerce packages, but some require more customization than others. This feature is important in business-to-business commerce where most companies work with PO instead of credit card transactions. One of the weak links of the typical PO system is automating the updating process of internal systems between companies.

Although the majority of payments at Web-based stores are made through traditional credit card transactions, you need to be in position to accept as many payment methods as possible in order to maximize your sales. The most common credit cards accepted by these packages are Master-Card, Visa, American Express, and Novus/Discover. However, you can also customize a package to include other credit cards. On the credit verification system side, Cybercash and VeriFone represent the two vendors that usually provide easiest integration into commerce software. Of the two systems, Cybercash dominates the market. Other payment methods include micropayments, smart cards, and online billing systems. Choose a solution that supports a wide range of electronic and traditional payment methods. For more information, see the Payment Systems chapter.

Most commerce packages include the option to specify sales tax and shipping costs. Calculating a customer's sales tax and shipping cost represent vital components of any good e-commerce software package. Customers want to know their total cost prior to placing their order. Shipping costs need to be calculated in real time. Further complicating this issue is the notion that a significant proportion of orders in your Web store may come from international customers. Select a solution that allows you to calculate and specify shipping costs with a great amount of detail.

Remember that taxes vary by state and by nation. How do you address this element of your commerce package? Third-party tax and accounting packages such as Taxware International or Great Plains Accounting Software allow you to integrate their product into your Web site. Even some of the store design wizards support this integration.

Site analysis provides you with the option to track visitors to your site, products purchased, and pages accessed. Since several of these packages have thousands of items you can analyze, we recommend spending some time planning prior to setting up the tools. Decide exactly what you want to monitor. If you do not plan out exactly what type of trends and analysis you are trying to capture, the analysis tools available could overwhelm your efforts. The goal is to produce valuable information, not just data. And for all you technologists reading the book, ask the marketing, sales and product development departments what information is valuable to them for their work.

Commerce packages vary on how much they offer for site analysis. Those packages that do not provide the features you need typically have third-party solutions that you can substitute. It is important that you review site analysis results to determine which advertising campaigns are generating sales, which pages are not accessed, and finally, how effectively your site supports your e-commerce ambitions. You can use this information to customize your site for an optimal return on your Web site investment.

Customer-Oriented Features

The following section addresses commerce software features from the perspective of the customer. These features include product listing, discount pricing/auction technology, question and answer listings, product comparisons, order tracking, shipment tracking, and search tools.

Some software packages allow you to place products in more than one category. By classifying products in multiple categories, you ease your customer's shopping experience. This is a compelling feature for two reasons: products get more exposure because of their availability in multiple categories and products become easier to locate. Planning your product categories is critical. You do not want to have too many or too few categories. Ensure that you understand how a potential customer views the product information. Consider usability testing of your site.

Discount pricing implies a product sale. Commerce software allows the discounting of selected products, "two for one" sales, and sales based on the amount purchased. Some software allows you to address specific customers for discounts. For example, you may have preferred customers or specific buyers may employ different pricing schedules, based on negotiated agreements.

Auction technology software represents a common plug-in for several commerce vendors. Auction technology harnesses the current product information in your storefront and allows you to select the items you would like to auction. Most of the auction software plug-ins that are part of commerce packages require a considerable amount of customization. However, if auction technology is a central element of your business processes, its deployment is worth the time and effort.

Question and answer listings create the impression and added value of a sales clerk working at your Web site. You assemble a series of questions that you anticipate your potential customer might ask and add them to your site. When your customer needs information about a product or service, he selects and responds to the appropriate question.

The response leads the customer through additional questions until the product or service has been selected.

Perhaps an example would best illustrate the process. Tommy enters your site with intentions of purchasing a bike online. The page presents a series of questions to help determine the appropriate bicycle. What type of bike are you looking for: mountain bike, racing bike, or street bike? Once Tommy selects the mountain bike category, the next question narrows Tommy's budget. $100-150? $151-200? $201-250? $251 and over? The next question concerns vendor preferences. After determining that Tommy does not have a preference among vendors, the system displays pictures of bicycles that fit the selected criteria, their specifications, and a price list. Tommy can then ask for more information or decide which bicycle to buy.

Product comparisons allow your customer to compare features of two or more related products. Prior to implementing this feature, ensure that your products or services justify this type of information. For example, if your customer wanted to buy a car, they would be very interested in comparing features such as miles per gallon, safety features, etc. Product comparisons would make sense in this case. However, you are not adding value to your customer experience if you attempt to compare products that are unrelated (for example, a blender with a hair dryer).

Another standard feature of commerce software is order tracking, both internally for your company and externally, for the consumer. Order tracking allows you to view pending, disapproved, and processed (completed) orders. Typically, once a customer's credit card is approved, the status of an order changes automatically from pending to approved. For purchase orders, the process may include a combination of automated approvals and manual approvals. The goal, however, is total automation of the process. Once an order is approved, the order is completed, and the product is shipped to the customer.

Most commerce packages provide an option to allow customers to track their packages via UPS or Federal Express (FedEx). For example, tracking is part of UPS Internet Tools, which is designed to help merchants serve their customers better. UPS Internet Tools allows customers to track their packages worldwide without leaving your commerce site. UPS tracking is available in local languages in 16 countries and territories via the Internet. Tracking is also available via the UPS home page.

FedEx, the other dominant shipper, is now offering a set of tools to provide shipping and tracking functionality. FedEx ShipAPI provides you with a shipping application programming interface (API), sample templates, and instructions on how to connect seamlessly to FedEx. It streamlines your online shipping process by integrating FedEx shipping templates and tools into your company's site or corporate information system. You can connect directly to FedEx when placing shipping orders and scheduling pickup requests.

FedEx TrackAPI provides you with a tracking application programming interface (API), sample templates, and instructions on how to connect seamlessly to FedEx. Enable users to check on the progress of a shipment instantly from their own Internet-enabled computer. Track multiple packages simultaneously, consolidating several inquiries into one simple transaction. Access available shipment information, using ship date and country code or account number to check progress. Customize sample templates for tighter integration into your online environment. Retrieve and store additional shipment information to provide your users instant access to detailed historical shipping data. Built-in error-checking capabilities verify the accuracy of required tracking information. The tracking system uses public-key encryption software to help protect confidential data information, as well as verify data integrity and authentication.

As the popularity of the World Wide Web gains momentum, the amount of information that is available online rises dramatically. Site visitors experience increasing difficulty in finding specific information. If site visitors can not locate the information they want (and quickly), they move on to another site. Commerce solutions provide site administrators with tools to gather and index information and to create keyword search capabilities on your site.

Keyword search capabilities enable site visitors to use highly targeted search queries to find and retrieve the specific information they need. You can gather and index information by crawling the Internet, an intranet, Exchange Public folders, NNTP newsgroups, an ODBC database, or a file system. Document properties are the characteristics of the document, including the author's name, date created, title, subject, documents address, and contents of the document. Once extracted, the document properties are stored in a central location.

In addition, search tools provide your site visitors with a Web page from which they can initiate a search, and a page to display the results of the search. These pages can be extensively tailored to the needs of your site visitors and to the information that is available in your catalogs.

Advertising Features

The following section addresses commerce software features from the perspective of your company's marketing efforts. These features include registration, cross-selling, advertising and promotion tools, push technologies, and mail distribution tools.

Customer registration is a standard feature for commerce software solutions. Registration may be set up as soon as visitors access the site, as soon as visitors want to purchase a product or service, or when visitors access confidential information. For the first scenario, where access to information is dependent on the user login name and password, multiple vendors can access your site and be provided different products or services to purchase. An example of the second scenario is the Wall Street Journal which requires that you register and pay a fee when accessing the newspaper online. An example of the third scenario is American Express which offers online options to view credit card information. The site requires that visitors enter information prior to accessing their personal accounts.

Another reason why registering visitors is extremely important to your company is that it allows you to capture data about your customer. This data can later be used to analyze buying trends, send product information back to your customer, and set up personalization of options based on their preferences.

Cross-selling features use shopper trends and existing transaction information to recommend products of interest to customers. For example, Amazon recommends additional books to customers based on their previous purchases or based on a book they are browsing.

Think about visiting a store. The natural links between products arise almost automatically. When you visit a camera store, the store owner strategically places all of the other associated items that you may need: film, batteries, flash, photo albums, etc. By the time you finally leave the store, you have spent twice what you originally intended. As a store owner, offer products associated with the products that your customer seeks. You provide convenience. The customer provides additional revenue.

Once your store is operational, you need the option to run promotions. You should be able to quickly change a price or offer discounts for volume purchases or to special customers. Advertising capabilities may include banners, featured products, special discounts, etc. Some of these packages even provide options to track how many clicks have been made on each banner and allow you to set up the banners and ads so if a visitor searches for a specific piece of information, a related banner displays. Other types of ads included are "buy now" ads for impulse buyers. Some commerce software includes great options for advertising. Other commerce software offers very little, requiring that you purchase a third-party advertising plug-in or build the functionality into the site.

Push technology provide users with channels that deliver specific content. The user subscribes to a channel and related information is pushed (delivered) to the browser on the desktop. Most people have heard of PointCast, one of the first push technologies out on the market. Now commerce vendors provide solutions featuring push technology as one of their delivery mechanisms.

Direct mail is similar to an electronic version of paper-based direct mail. The difference is that you can use direct mail to deliver personalized content to thousands of users with lower costs than paper-based mail. In addition, direct mail offers automated delivery (a standard of marketing).

Server Management Tools

Commerce server management tools may be included as part of the commerce software or as part of the management software provided with a commerce solution. Most vendors include the following capabilities: content management, replication and clustering, site usage statistics, and remote administration.

Content Management

Content management is a feature which allows you to manage your content, including new products, the design of pages, and additional categories. Think of content as all of the information displayed on the site prior to that information being displayed to the world. Content management is critical to the success of any site. The world is full of visually stunning sites. However, sites often feature misspelled words, poor grammar, incorrect data, etc. Remember that your customer's first impression could become a last(ing) impression.

When selecting a commerce package, ensure its content management capabilities. This includes the ability to review your information prior to replicating it to the servers that your customers or suppliers access. Typically, content management involves a central hub server where all changes are made. Approvals are set up by assigning privileges to specific employees within the company. These employees are required to approve the pages prior to sending them out to the external servers. If they are not on the list as an approver of the information, they do not have the right to modify the information. Once approved, the pages are replicated out to the production servers. This review should include both the Web page design and the specific contents.

Figure 6.1 illustrates a typical content management process. The author of a document sends (1) the document to the hub server, the person responsible for authorizing the document reviews (2) the document needed to be approved. Once approved, the document is sent (3) to the external commerce servers. Clients retrieve (4) the document from the commerce server.

Figure 6.1 Content Management

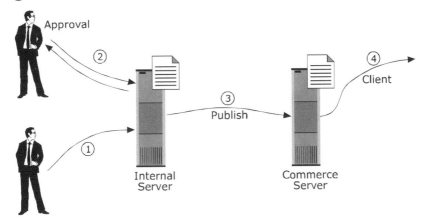

Replication and Clustering

Replication and clustering capabilities are standard features of the commerce software or management component. Replication and clustering helps you scale your servers. Scalability is critical when you have more and more servers that Internet visitors are accessing. Instead of manually transferring files between servers using FTP, use the replication capabilities provided by the commerce solution. Even the smallest sites should use this feature, so that administrators do not have to manually copy files between servers.

Site Usage

In order to determine when to scale your site, employ site usage tools to provide you with a mechanism to track server performance. Performance ratings are based on the number of processes running on a server, the number of users accessing the site, the hits per server, CPU usage, and memory usage. These commerce programs provide the tools needed to ensure that your servers perform efficiently. You do not want a system that serves Web pages to visitors slowly. If the site is at 70% utilization, add another server.

Remote Administration

Remote administration of the server also represents a standard capability in a commerce package and/or server management package. Do not manage your site on the same computer that your customers are accessing. The servers should be in a secured location with limited access. Accidents, such as deleting or modifying the wrong files, or even spilling coffee during those late nights on the machine, are another compelling reason for remote administration. In addition, you do not want to drain additional computer resources from the servers that your customers access.

With remote administration, you can modify security settings for your network operating system, Web server, commerce server, or almost any other software on the machine. This feature allows you to view all of the servers in your environment from one central console. You can select one server and make all the changes necessary, select the next and make the changes, etc. You can then view the information as if you were sitting at the remote server.

Integration

Linking your existing inventory management and finance systems to your commerce software remains one of the most challenging aspects of setting up a successful e-commerce site. Before selecting a commerce server software package, make sure that you can easily integrate it with your existing back-end systems. Make sure you know exactly what systems you have in place. Your commerce solution requires that you pull data from these systems to your commerce site. Gather the names of the software vendor and the software packages with version number. In addition, you need to understand your current infrastructure.

Study the role of the following systems within your company's infrastructure: network operating systems, Web servers, relational databases, transaction-based systems, ERP systems, EDI systems, third-party software, and also any proprietary systems.

Network Operating Systems

The network operating system that supports your enterprise plays a integral role in your choice of commerce software packages. Some commerce software packages and integration tool sets, especially Microsoft's, require that you use Microsoft NT. Other vendors such as IBM and Lotus are very platform-neutral. The network operating system should not be the make or break factor when selecting the commerce software. However, if you are a UNIX-based shop and are moving to a Microsoft solution, realize that such a deployment requires new hardware, software, and skill sets for an effective implementation.

Web Server

Most companies already use Web servers. And while the version(s) of Web servers you employ influences your commerce solution, the version does not dictate a specific solution. Several commerce packages support the use of a variety of Web servers. For example, IBM's commerce solution allows you to use Netscape's Web server, Lotus' Web server, and other brands of servers.

If you decide to deploy commerce servers from other vendors, you can still connect the commerce and Web servers by using links or URL addresses. For example, IBM uses a variety of Web servers at their site. They have seamlessly integrated the site among the contents on the Web servers. Their site took advantage of the technologies that best met their business needs.

Relational Databases

Integration with existing relational databases is provided in most commerce package solutions. Each vendor has its own set of tools that provides integration to specific relational databases. The question you must answer is how easily a commerce system can be integrated. Oracle seems to be the one database for which commerce vendors have provided easily integrated solutions. Other commonly supported databases include Microsoft's SQL and IBM's DB2. Even if there is not an out of the box solution, other database vendors, such as Sybase, have built tool sets that you can use to tie into these systems. Some database vendors have code at their site that you can use as a starting point for tying into these relational databases. Begin your research at the commerce system vendor's site to locate a tool to support database integration.

Transaction-Based Systems

The goals of the integration of commerce systems to transaction-based systems are seamless processing of orders and timely, accurate updates. Most commerce vendors offer connections to a variety of transaction-based systems, although with varying degrees of compatibility. The questions you must address in your research include how difficult is the implementation and how robust is the solution.

The focus of IBM and Lotus includes integration into any back-end environment, especially legacy platforms (CICS and IMS) and transactional middleware (MQSeries). Their solutions have grown more robust and user-friendly with each release. Integration tools include database tables, scripts, sample code, and documentation to support transactions and message queues with other environments. In contrast, some vendors, including Microsoft, have not focused on providing the same degree of connectivity.

Enterprise Resource Planning Systems

Most vendors provide solutions to integrate with some of the most common enterprise resource planning (ERP) systems, including PeopleSoft and SAP. Commerce server systems add value to your existing ERP system by passing only fully formatted orders, by acting as a firewall, by eliminating direct access by Internet users, and by enhancing the catalogs presented to customers with more interactivity.

Third-Party Systems

If you have an accounting package, tax software package, or any other type of system that you want to integrate into the commerce environment, visit the commerce vendors and the third-party vendors to determine if an integration tool is available. If it is not, you need to write the API to perform the integration.

Electronic Data Interchange (EDI)

Electronic data interchange (EDI) allows program-to-program data exchange over private or public networks. It is based on a set of predefined relationships and standards between the participating parties. These predefined relationships and standards include interfaces, translators, consistent data, and mapping protocols. Commerce server solutions support EDI by delivering merchant orders or customer special orders to the supplier, and by allowing suppliers to report order status to the merchant.

Proprietary systems

Connecting a commerce system to a proprietary software system requires building an application programming interface (API). An API serves as a middleman by making requests between the two systems. For example, many companies use a proprietary credit card verification system. An API connects the commerce server to that proprietary system, ensuring efficient and accurate communication.

Customization

The typical commerce software implementation generates a storefront devoted to selling either products or services. The implementation incorporates a generic set of workflows. The graphical user interface is driven by the best guesses of its developers. However...

- You might want to offer your customers products and services.
- Your organization may use an entirely different set of workflows.
- You might not appreciate the resulting user interface.
- You might want to alter both the front-end and the back-end.

In order to vary from the existing configuration, you must programmatically reconfigure either the workflows or the graphical user interface, or both.

The Role of Wizards

The easiest way to configure a commerce server is to work through the series of installation and configuration wizards that dictate your commerce options. These wizards create thousands of lines of code. The resulting data entry fields are often related across several screens. Information on one Web page can affect information on another Web page. Although the tight integration of these components is in one sense to be admired, the downside is reduced customization.

How Much Work is Required?

Software, in this case represented by your commerce server, is composed of individual modules. Two measures of modules are critical in determining the effort required to customize your commerce server implementation: cohesion and coupling. Cohesion refers to the degree to which a program module's instructions are functionally related. A highly cohesive module implies that all the instructions within that module are working toward a common goal, a specific function. Coupling refers to the level of dependency that exists between the modules of a software program. Well-constructed modules are independent from each other, such that one module can be modified without dramatically affecting another module. The measure of cohesiveness and coupling for a module that requires modification indicates the necessary level of effort.

Modifying the Workflow and Business Processes

Customizing the workflows and business processes employed within your commerce software requires an expert programmer that is familiar with the package's development environment. The programmer should have the ability to read the existing code and to modify that code to adjust to your workflows or business processes. The more cohesive and less coupled a program module, the easier it is to modify.

Modifying the Interface

After you create a storefront using a wizard, you may want to make changes to its look and feel or to its navigational elements. Some vendors have created solutions that are very simple to modify. Even an entry-level Web designer could make the necessary modifications. Other vendors have made it almost impossible to alter the storefront, due to its complicated features. However, most major commerce software vendors provide sophisticated interfaces that allow you to get a store up and running quickly and painlessly. Your interface goal is to create an experience common enough to ensure your customers' comfort, yet unique enough to set yourself apart from your competition.

Skills

Now that the features of the commerce software have been explained, we turn our attention to the skills necessary to effectively configure the commerce server software. The skills needed for implementation are technical in nature, and include knowledge of network operating systems, back-end systems, Web servers, programming, and graphical user interfaces. Who is qualified to do the work? Obviously, only a team would have the entire spectrum of skills described. After taking an inventory of your team's resources, you are faced with a decision, "who is going to develop the solution?"

Internal development of the commerce server software implies that you will use existing resources, either current staff members or new hires, to implement your solution. An important consideration when choosing this path is whether the solution requires continual tweaking. That is, how much maintenance effort is necessary to keep the commerce server implementation current? The advantages of configuring commerce server software internally include a lower total cost, if the team has the appropriate knowledge and skills, and a stronger sense of ownership. Internal development ensures a familiarity with the configuration. The potential disadvantages include a longer development time, political conflicts within the organization, and missed opportunities if developers are unaware of recent technical innovations. Remember, if you favor this option, your company supports the project from start to finish. A careful analysis of your team's skills often leads to a solid decision.

External development of the commerce server software implies using outside resources (typically, consultants, technical support providers, or other vendors) to implement the solution. An important consideration when choosing this option is whether the solution is a one time effort. That is, are most of the development and implementation efforts accomplished at the start of a project? Advantages include opportunities to use best practices, to focus on your current business instead of the implementation details, and potentially a quicker delivery. Disadvantages of using external resources to develop the software solution may include a higher cost and a lack of leverage with outsider's exposure to your business process knowledge.

At the heart of an e-commerce solution lies the commerce software, with capabilities for creating an online storefront. Several major vendors on the market offer a variety of software solutions that allows you to establish an e-commerce site on the Web. The components of this software include storefront implementation tools, commerce server management tools, and back-end integration tools. In order to select among these solutions, first you must understand how your company intends to participate in the e-commerce marketplace. Then you must understand the skills of your team members. As you gather information about commerce software vendors, consider how these products blend into complete vendor solutions.

7

Vendor Solutions

- Reviewing vendor solutions
- Evaluating e-commerce packages
- Implementing vendor solutions
- Buying or outsourcing services
- Hosting the solution

Software vendors provide a variety of compelling solutions for both business-to-business and business-to-consumer e-commerce. Most of these solutions incorporate the basic storefront functionality necessary to sell products and services, and management tools for controlling content and performing searches. Other vendors provide more robust capabilities in areas such as site analysis, marketing, and advertising. As you consider the vendor solutions described in this chapter, remember that your focus is on the needs of your business.

So which vendors provide compelling e-commerce solutions and how do you select among them? On the market today hundreds of vendors offer e-commerce solutions. To provide a starting point for your solution, we focus on five of the top vendors. However, there are other solutions to choose from that may also meet your business needs. In some cases they may cost a little less than those covered in this chapter. The dominant e-commerce solution providers include Microsoft, IBM, BroadVision, OpenMarket, and INTERSHOP.

This chapter also explains the attributes to consider when you evaluate any e-commerce package. Important attributes include software features and functionality, commerce server management tools, back-end integration tools, and additional plug-ins provided by the e-commerce vendor or third-party vendors. Cost obviously plays a role in your final decision. However, it should not be the main factor when selecting a software package. If the software package offers a majority of the features and functionality you require for your site, it may prove more reasonable in the long run. Imagine having to hire two developers for a couple of months or more to create (with new code) the exact same features and functionality for your storefront that was featured in the "more expensive" solution.

A working knowledge of your options is critical to your success. If you intend for a commerce service provider to host your solution, ensure that the company satisfies all of your requirements. Commerce service providers represent different vendor solutions. For example, some only host Microsoft's solution, while others may only host IBM's commerce solution. We have included a discussion of commerce hosting services for each vendor surveyed in the chapter.

Evaluation Guidelines

Evaluation guidelines offer a comprehensive way to evaluate and to compare each vendor software package. Each vendor solution section reflects the evaluation guidelines established at the outset of this chapter. Guidelines for evaluating each solution include an overview of features and functionality, as well as descriptions of commerce server management tools, back-end integration tools, vendor and third-party plug-ins, and hosting services.

Another great way to evaluate a vendor solution is to look at their customers' profiles and determine how they successfully implemented the technology. If other companies are successfully employing the vendor solution, you will not become a lab rat, serving as a test for the vendor. In addition, customer solutions help you become familiar with the overall architecture, design, and scalability of a vendor's software and hardware solution.

Several research companies on the Internet provide up-to-date reviews of e-commerce packages. These sites are definitely worth investigating. For example, www.e-comm.internet.com has a section dedicated to reviewing vendor software packages available on the market. The site rates software packages with one to four stars and provides a brief description of each package's pros and cons. Other research company sites, such as www.sellitontheweb.com and www.internet.com, have complete reviews of e-commerce packages from both small and large vendors.

We often encounter companies that delay their purchase decisions while a solution vendor tweaks a new release with the latest and greatest features. The next release offers X, Y, and Z functionality. Now is not the time to delay. Your goal is to design, develop, and deploy a compelling solution. Once your company is online, you can add more features as they become available.

Features

Each major vendor solution includes a variety of features, although the quality of their implementation varies. Devote some effort to distinguishing between vendors and to learning their strengths and weaknesses. The main features found in a commerce software package include the core electronic commerce functions that help you to get your site up and running as quickly as possible, and at lower expense. Based on your business needs, ask the following questions:

- Does this vendor's solution allow you to sell products, services, and digital content?
- Does it provide flexibility in adding data fields or implementing new business processes?
- What type of advertising and promotional features does it provide? Banner advertisements, cross-selling capabilities, featured products, and product discounts fall into this category.
- Does the solution offer push technology features?

Additional questions about the solution include:

- Does the solution offer personalization for visitors?
- Does it provide capabilities for sending e-mail to specific customers or to all customers?
- Does the solution provide robust security and workflow approvals?

Commerce Server Management Tools

Commerce server management tools enhance the management of your Web site content, provide load balancing, include analysis of site usage, and enhance server performance. These tools allow you to easily scale your solution to keep pace with the growth of your business. Based on your business needs, ask the following questions:

- How does the software enhance content management prior to publishing the information on the Web?
- Does it provide replication and clustering capabilities, so that you can easily scale the system from one computer to several, if necessary?
- Does the solution incorporate an analysis tool to generate site usage reports? You need to know when to increase the performance of server hardware and software and when to add additional servers.
- Can you manage the software remotely?

Integration Tools

Back-end integration tools allow you to connect your databases to your e-commerce solution. To make a reasonable decision when selecting one vendor's product over another, you need to identify and define exactly what back-end systems you plan to integrate into the commerce environment. You also should document both the systems you have in-house and your business partner's system. Armed with this information, you can easily evaluate each vendor's product to determine how difficult an integration process you are facing. Be very critical of a vendor solution that does not allow you to integrate it with your existing back-end systems. Based on your business needs across the various components of your infrastructure, ask the following questions:

- What network operating system (NOS) is required?
- Which Web server is required to run the chosen commerce software? Some vendors allow you to run their commerce software on several types of Web servers available on the market.
- Does the vendor provide plug-in tools to the relational databases currently deployed at my or my business partner's company? Relational database systems supported by these vendors include Oracle, SQL, and DB2.
- Does the solution have the capability to integrate into my transaction-based systems (often, CICS and IMS)?
- Are there tools that provide connectivity to the enterprise resource planning (ERP) systems currently used by my organization? ERP systems include SAP, PeopleSoft, Baan, and JD Edwards.
- If you employ electronic data interchange (EDI), what connectivity is provided by the vendor solution?

Also, determine if plug-ins are available to any other third-party software solution you may currently employ. These software systems include document management software, image management software, and accounting packages.

Vendor Plug-ins

Companies such as IBM and Microsoft provide additional plug-ins for their e-commerce software packages. For example, both IBM and Microsoft sell an auction technology plug-in. Other commerce vendors provide plug-ins for package tracking through UPS or FedEx, additional management tools, and development tools. This information is usually provided at the vendor's Web site. Since these vendors continue to improve their products via plug-ins, it proves very advantageous for you to keep updated. Visit vendor sites at least every couple of weeks to determine what additional tools have been developed to integrate into your e-commerce solution. Remember that it is much easier and faster to deploy a plug-in, than to create the same functionality.

Third-Party Plug-ins

Third-party plug-ins parallel those of the major vendors, but are offered by other companies. Hundreds of independent software vendors (ISV) provide direct plug-ins. Find out what is available to help you evaluate your potential commerce solution. Again, remember that it is much easier and faster to deploy a plug-in, than to create the same functionality from scratch. Based on your business needs, ask the following questions:

- Does the solution provide the capability to enter payment system information for vendors such as CyberCash and VeriFone?
- Can I integrate my existing tax and accounting systems?
- Who are the vendors that provide plug-ins for procurement, bill payment and presentment, inventory tracking, and UPS/FedEx tracking?

Commerce Hosting Services

For those organizations that do not have the resources to host an entire e-commerce solution on their own site, commerce vendors provide a reasonable alternative. Companies, such as OpenMarket, IBM, and Microsoft, have developed commerce-based products specifically for Internet Service Providers that want to host their customer's commerce-based solutions. Depending on the vendor software solution you select, you can choose from a variety of ISPs to host the entire server system and commerce software for you. Each vendor site includes information about their commerce hosting services.

Customer Profiles

Who are the customers and how are they using vendor solutions? The reason we ask these questions is to avoid being treated as a testing ground. We do not want to serve as a learning opportunity for a vendor. If a vendor has success- fully deployed their solution, your confidence is buoyed. In addition, a competent vendor can support your installation with best practices (developed under another company's watch). The vendor's consulting divisions or business part- ners have the experience necessary to help your company with the planning and design of your e-commerce infra- structure and application. You do not have to start from scratch and learn what they already know.

In addition, these vendors provide case studies of show- case customers and the technologies used to deploy the com- merce solution. By reviewing the case studies, you may learn about systems your company has in house that have been integrated into the vendors' technology. For example, Microsoft and Lotus Domino do not offer a tool for back-end integration to Great Plains Accounting Software. However, after reading a case study from these vendors, we noted that this accounting package had been integrated into the archi- tecture. We visited Great Plains Web site and learned that they had built a tool that integrated into both platforms.

Let's start the vendor evaluations by looking at some of the major e-commerce vendors in the areas of major fea- tures and functionality, integration tools, third-party plug- ins, and hosting services. This will provide you with a start- ing point to research the best solution for your company. We highly recommend that you access vendor Web sites to get the full details of each package.

Microsoft

Microsoft represents one of the top vendors of commerce server software. Companies such as 1-800-Flowers, Dell, Barnes & Noble, and Eddie Bauer have used Microsoft's solutions for selling products and services to customers on the Internet. Like most commerce software vendors, Microsoft's Commerce Edition software provides an out of the box solution to easily create a storefront used to display products, manage customer orders, and register users online. In addition, the software incorporates compelling features such as cross-selling, group discounts, and advertising banners.

Features and Management Tools

Microsoft's Commerce Server provides basic storefront capabilities. The software allows you to add, delete, and edit product information, place products in promotions, calculate product tax, calculate shipping and handling charges, and manage customer orders. Pricing information for Microsoft's latest Commerce Server software program can be reviewed at www.microsoft.com, by entering the Commerce Server product information section.

Microsoft also provides an advertisement server used to place banner ads (your ads or a partnering company's ads) on your site. The software includes options to specify the number of clicks for each banner ad, the scheduling of the ad, and the analysis of how well the ad performed versus the projections. The server is sophisticated enough to display a specific banner that has been associated with the keyword a user enters. By linking keywords to specific banners, you can display a mortgage company banner advertisement when a visitor enters the keywords "real estate."

In order to help you manage the site, Microsoft developed a product called Site Server. This product provides basic management features, such as content management and site analysis. The product also has the capability to deliver targeted information to your customer via push technology and to send e-mail to targeted groups via a direct mailer. Direct Mailer is an easy-to-use tool for creating a personalized, direct e-mail marketing campaign based on Web visitor profiles and preferences.

Integration Tools

Microsoft and other third-party vendors offer several tool sets for integration into back-end systems. A minimal requirement in this architecture is Microsoft's SQL or Oracle's database system to store product and customer information. Microsoft's solution also requires Microsoft's IIS server as the Web server. Tools are also provided to connect to IBM's CICS and MQSeries. Other vendors such as SAP, IBM, and Great Plains Accounting, have created their own tool sets to allow seamless integration into Microsoft's commerce software. Keep in mind that vendors are building more and more tool sets to integrate into Microsoft's platform because Microsoft represents one of the top vendors in the e-commerce software market. If you have a system in your company that is not on Microsoft's list, request that the vendor build a tool set for integration. A positive response could save you a lot of time and money.

Microsoft and Third-Party Plug-ins

Microsoft provides an auction component which is fairly simple to incorporate into your commerce site. However, Microsoft relies heavily on independent software vendors (ISV) to provide other plug-ins for their commerce solution. ISVs such as CyberCash, VeriFone, and CyberSource provide payment solutions that easily integrate into the commerce site. Other vendors such as TaxWare, Pandesic, and Tellan offer tax calculation software. More and more vendors are focusing on offering solutions that easily integrate into the Microsoft solution. Go to the Microsoft site to get a listing of ISVs and the solutions they provide. Expect this list to continue to grow.

Commerce Hosting Services

Several companies host Microsoft-based e-commerce solutions. To find a listing of commerce host providers, go to www.microsoft.com and search for commerce host providers. The results include an updated list of providers. Commerce hosting partners for small-to-medium businesses have worked with Microsoft in order to create a complete commerce site solution for a low setup price and monthly rate. Commerce hosting partners for medium-to-large enterprise businesses have worked with Microsoft to offer higher-end solutions for larger corporations.

Customers

A wide variety of companies successfully employ Microsoft's e-commerce solution. Barnes & Noble used Microsoft products to quickly create an easy-to-access electronic storefront. The frequently visited CBS SportsLine site has converted a subset of its existing back-end commerce applications to the Microsoft commerce platform. Cooking.com used Microsoft Site Server Commerce Edition to quickly create a Web site featuring dynamic content. Dell Computer Corporation uses Microsoft technology to manage its rapidly growing electronic commerce business. Casual lifestyle product retailer Eddie Bauer developed a site that runs on the Microsoft commerce platform. Using Microsoft e-commerce software, Intell-A-Check enables Web-based businesses to provide online customers with a secure alternative to credit cards. To lower costs, Mobil Corporation is piloting the TransPoint Internet-based bill delivery and payment system. The Motley Fool runs its entire investing business on a Microsoft commerce platform. Nautica Apparel, Inc. gives its business partners 24-hour access to critical information with a new intranet/extranet solution based on the Microsoft BackOffice family of products. Remember the breadth and depth of these success stories when considering Microsoft as a candidate solution.

A New Direction

This year Microsoft announced a new direction for their commerce software efforts. The proposed solution is based on a system they call BizTalk Framework for e-commerce. The goal is to develop a software solution that easily integrates into any application. The BizTalk framework is based on new Extensible Markup Language (XML) schemas and industry standards that enable integration across industries and between business systems, regardless of platform, operating system, or underlying technology.

Microsoft is working with customers such as 1-800-FLOWERS, BarnesandNoble.com, Best Buy Company Inc., Dell Computer Corporation, J.D. Edwards & Co., PeopleSoft

Inc., SAP AG, and other customers and industry vendors to define BizTalk schemas. These schemas describe common business processes for exchanging information and documents between trading companies. By defining a technical vocabulary for describing common business processes in electronic commerce and across specific industries, information can be easily exchanged between companies, regardless of the platform.

IBM

IBM is also one of the leaders in e-commerce software solutions. This year they have emphasized selling their e-commerce product line called Net.Commerce. This commerce software is aimed at both business-to-business and business-to-consumer commerce. Emphasis has also been placed on providing excellent product catalog features and seamless integration with most back-end systems. IBM's market share is continuing to grow worldwide as they help companies by defining their strategic commerce direction and the software required to implement the solution. This section surveys the highlights of IBM's product line.

Companies such as 1-800-Batteries, MicroAge, UPS, and Macys have implemented Net.Commerce as their e-commerce software solution. Other companies such as Daimler-Chrysler, have used a combination of IBM solution and Domino, a Lotus-based product. To review some of the other e-commerce solutions implemented using IBM's technology, go to www.ibm.com and access the e-commerce section of the site.

Features and Management Tools

IBM offers two versions of their software package, Net.Commerce Start and Net.Commerce Pro. Both products generate storefronts to display products and services, manage orders, and track inventory. Other common features include tax and shipping cost calculation and a variety of payment options. Some of the advertising features include cross-selling and group discounting. Applications that have been integrated into the solution include Seagate's Crystal Reports which generates statistical reports and Net Perceptions which helps sellers make real-time predictions about individual customer preferences. IBM also offers a credit card payment feature called IBM CommercePOINT eTill that easily integrates into the commerce solution.

IBM Net.Commerce Pro includes some capabilities to personalize a customer's onsite shopping experience. The component, called the Advanced Catalog Tool, provides a variety of compelling features. Intelligent searching quickly narrows the search for products and eliminates those that do not meet specified parameters. Sales assistance presents a question and answer selection process to narrow down the range of product choices for each customer. This feature offers the customers a personalized shopping experience by directing them to the requested product. Finally, product comparisons allow the customer to compare several products they have selected. The products are then placed side-by-side on the screen and the different features are highlighted so users can easily compare them.

Integration Tools

IBM provides tools to easily integrate into several major ERP, transaction-based, and relational database back-end systems. A minimal requirement is that the product and customer data are stored in a relational database. Out of the box, IBM's solution is very simple to integrate with DB2 and Oracle. However, you can also integrate with most relational databases. In addition, IBM can run on several Web servers, such as Domino, Domino.go and Netscape servers.

Net.Commerce Pro has additional tool sets to integrate with EDI, IBM CICS, MQSeries, MS products, and SAP. Other vendors are also building solutions to integrate with IBM's solution. Again, prior to building the integration tool, check with the vendor to determine whether they have already built it.

IBM and Third-Party Plug-ins

IBM and many ISVs provide compelling applications that can easily integrate with the commerce solution. For example, IBM has an auction component that easily plugs into Net.Commerce. In addition, you can include UPS tracking for customers to track the status of their shipments. Other vendors such as Assist Cornerstone Technologies provide software ranging from order entry, inventory, purchasing and sales analysis to general ledger and accounts receivable financial accounting and analysis. CyberSource, CyberCash, ICVERIFY and VeriFone have integrated credit card solutions. In the advertising and analysis area, net.Genesis (www.netgen.com) offers a tool to capture detailed information about the user traffic, including what is effective, what is ineffective, and provides an overall understanding of the entire Web site. E-centives (www.e-centives.com) provides digital coupon technology. Narrative Communications (www.narrative.com) and NetGravity (www.netgravity.com) provide tools to integrate online advertising and direct marketing solutions. Many more vendors offer plug-in solutions. To find out which vendors provide plug-ins, go to the IBM site and select the ISV section.

Commerce Hosting Services

IBM offers a special edition of Net.Commerce for hosting e-commerce sites. There are several ISVs that use this solution to host commerce sites for small, medium, or large companies. For example, Internet Tradeline provides hosting of IBM commerce solutions (www.tradeline.net). In order to locate these hosting companies, go to the IBM Web site and look for the commerce hosting providers.

Customers

A wide variety of companies successfully employ IBM's e-commerce solution. As a small company, Best-of-Italy was able to take giant strides towards becoming an e-business by linking its suppliers to its customers through an online mall, and by guaranteeing customers secure electronic ordering and bill payment. The Chrysler enterprise reengineered its supply chain management process cost-efficiently with a strategic combination of legacy systems and new technology. Egmont moved its catalog business to the Internet with an online store and a real-time stock inventory system, creating an end-to-end e-commerce solution that has streamlined the company and improved customer service. Lister Petter, a major manufacturer of industrial engines, made the leap to e-business by linking its worldwide distributors and automating the ordering and delivery of parts. MEXX, a large European wholesaler/retailer, improved customer service and reduced operating costs by leveraging its legacy systems into a Web-enabled supply chain management system. Office Depot of Mexico, investing in an IBM Web-enabled e-commerce solution, opened their company's catalog business to a new market and increased efficiency throughout the supply chain.

Transport, one of Canada's leading regional freight carriers, offered their customers tracking as a value-added service, in order to achieve a cost-effective e-business edge. Olan Mills improved both its internal operations and its ability to service customers by Web-enabling its unwieldy workflow process with fast and reliable IBM servers and Lotus software. 1-800-Batteries, a medium-sized company, used IBM's Net.Commerce to recharge an already established online business, and succeeded in expanding its customer base by providing better service. The Russell Corporation tied together its supply chain more efficiently by combining legacy systems with IBM's DB2 universal database and building a comprehensive inventory tracking system for its distributors. SAAB used IBM e-business hardware to efficiently network dealers to an automated inventory tracking system, and improved communication and data tracking without compromising systems already in use among the company's retailers. Consider the variety of implementations suggested by these success stories when considering IBM as a candidate solution.

BroadVision

The third vendor profiled in this chapter, BroadVision (www.broadvision.com), provides robust business-to-business, as well as business-to-consumer solutions. Their solutions focus on several industries as well as provide an e-commerce storefront package. The primary solutions support retail distribution, financial services, technology/manufacturing, telecommunications, and the travel industry.

Banks such as Argentaria Bank Group and Banco Santander are using BroadVision's Financial Services component to provide customers with banking services online. American Airlines has recently implemented BroadVision to offer personalized messaging, dynamically generating content based on customer preferences and profiles. Customers such as Audible and Blackwells have implemented retail distribution service for selling goods and services to online customers.

Features and Management Tools

BroadVision offers a product called One-To-One Commerce for storefront business-to-business and business-to-consumer sales of products and services. Some of the main features include user registration, order processing, product categorization, and search capabilities. The product also provides features for marketing and promoting items in the store. These features enhance cross-selling or up-selling to your customers, based on previous purchases, creating on-the-fly product comparisons and targeted coupons, and creating site-wide sales.

A unique component of the software is the customer service component that allows customers to send inquiries to the customer service representatives within the company, provides customers with real-time order status updates, sends order confirmation e-mails automatically to customers, and allows customers to update or change any part (or all) of their user profile.

Integration Tools

BroadVision commerce products use Oracle, Informix, Microsoft SQL Server, or Sybase to store product and customer information. The commerce software can also be run using Netscape Enterprise Server, Microsoft IIS or any CGI-compliant Web server. Additional back-end integration tools provide SAP, BAAN via e-mail, EDI, flat file, fax, or other data transport mechanisms. Other integrations supported by BroadVision include IBM MQSeries, Active Software, BEA, Tuxedo, and Tibco.

BroadVision and Third-Party Plug-ins

Several vendors also provide plug-in tools for BroadVision's commerce solution. These include shipping systems such as Federal Express and UPS, tax calculations using TaxWare and Tax Open software, and payment systems such as Veri-Fone and CyberCash. BroadVision also supports electronic software delivery through integration with CyberSource. Other vendors are also building solutions for BroadVision's commerce package. To review the latest list, go to www.broadvision.com and locate the Partners section.

Commerce Hosting Services

BroadVision plans to extend their market to commerce hosting services. Check with the company at their Web site for news of any progress in this area.

Customers

BroadVision's customers typically implement commerce solutions within the fields of retail distribution, financial services, technology/manufacturing, telecommunications, and travel. In the field of retail distribution, Audible, Blackwells, Fingerhut, and RS Components represent the success generated by a BroadVision deployment. Audible sells audio programming and merchandise online and takes advantage of cross-selling and up-selling opportunities. Blackwells integrated several different systems into a technically complex site renowned for customer service and the loyalty it inspires. Fingerhut built a robust and highly successful site to sell overstocked merchandise "dirt cheap" that integrates directly into the back-end inventory and shipping systems and is easily maintained by one non-technical person. And finally, RS Components offers customers another convenient and personalized choice for doing business with it online, through its fully transactional catalogue-based Web site.

BroadVision provides industry specific solutions for banks and financial institutions. One-To-One Financial offers deployment of personalized banking and online bill payment services. One-To-One Knowledge, for content publishing, management, and distribution, enables agents, brokers, and financial consultants to access and offer personalized knowledge online. One-To-One Enterprise supports customer relationship management, enabling financial institutions to increase customer retention, cross-selling, and up-selling. One customer of BroadVision, Argentaria Bank Group, offers more than 100 online banking transactions on its award-winning Web site (named best Internet application at COMDEX in 1997). The first Spanish bank to offer Internet banking services, Banco Santander, is rolling out new services every two weeks to their customers and achieving unprecedented growth. Bank Inter offered Internet banking to its customers in just 70 days from start to launch. And its daily online transaction volume exceeds that of any single branch. Liberty Financial also interacts with mutual fund customers on a one-to-one basis through an Internet fund management service for its SteinRoe mutual fund company.

BroadVision's technology and manufacturing solution focuses on supply chain management and larger scale e-commerce solutions with high transactions. The offerings include One-To-One Knowledge, One-To-One Commerce, and One-To-One. One-To-One Knowledge supports online content publishing, management, and distribution. One-To-One Commerce enhances deployment of personalized Internet transactions. BroadVision's One-To-One features personalized online relationship management with customers, partners, and employees across the extended enterprise. Another BroadVision customer, BAAN, spearheads just-in-time knowledge access for customers, partners, and employees, and shortens delivery cycles considerably.

BroadVision's telecommunication solution helps telecommunications companies manage customer interfaces such as operator services, directory assistance, billing, sales, and negotiation. Telus hosts Web services for small and medium-sized businesses and helps them target marketing efforts to individual customers online.

BroadVision's One-To-One travel solution is used to manage and integrate highly complex reservations systems and offers personalized, real-time travel planning in the burgeoning online travel industry. American Airlines created the world's largest and most advanced one-to-one travel Web site, enabling it to better meet the needs of its 31 million+ AAdvantage travel award program members. Thomas Cook created an online service to complement its established distribution channels. With BroadVision's One-to-One approach, Thomas Cook learns more about its customers, in order to serve their travel needs more effectively.

OpenMarket

Another dominant vendor of e-commerce solutions is Open-Market, named by Dataquest as the worldwide leader for Internet commerce software, with a 20% percent share in the global Internet commerce marketplace. OpenMarket's customers include Lycos, America Online, Consumers Union, Business Week, Standard & Poors, USA Today, Playboy Online, RealNetworks, Acer, Ingram Micro, Milacron, and Cablevision.

OpenMarket is also the leader in the commerce service provider market with a worldwide infrastructure of telecommunications providers, banks, Internet service providers and portals. Some of these companies include AT&T, Barclays Bank, First Union National Bank, France Telecom, Hiway, Netcom, NTT, Sage Networks, SBC Communications, Swiss PTT, Telecom Italia, and Telstra. OpenMarket has also provided several solutions for smaller businesses to get online at very little cost. OpenMarket's comprehensive Web site (www.openmarket.com) details additional customer solutions, ISVs, and hosting partners.

Features and Management Tools

OpenMarket separates its e-commerce products into two main categories by function: commercial Internet presence and order management. Each category provides very distinct features and functionality. We focus first on the commercial Internet presence solutions and then on their order management solutions.

OpenMarket provides three solutions based on the size of your business and whether you are selling products and services or digital content. These three options include LiveCommerce for business-to-business manufacturers or retailers, ShopSite for medium-to-small business storefronts, and Folio for electronic sales of information.

LiveCommerce is an Internet catalog-based application for creating search capabilities and for categorizing and maintaining thousands of products. Each catalog may be customized for specific companies, job titles, and individuals on the Web. This customizing may include search capabilities, and advertising and pricing specific products.

To include registration, authentication, order capture, fulfillment, payment, customer self-service, customer service, reporting, and analysis capabilities, OpenMarket provides a product called OpenMarket Transact (discussed later in this section).

ShopSite, for medium-to-small merchants, allows the merchant to set up a store in as little as 15 minutes. The store creation wizards provide all of the features and functionality to run a fully functional e-commerce solution. ShopSite includes features such as an "on sale" module, product up-selling and cross-selling, and associates tracking. The software also provides order management and processing. Using the Transact component, the software provides customer registration, tracks an employee's level of access within the store, offers self-service for customers (including a Smart Receipt and order status notification), and supports TaxWare and zone-based shipping.

Within ShopSite, there are several different versions based on your storefront requirements. These versions include ShopSite Lite, ShopSite Manager, and ShopSite Pro. Their features vary, depending on whether you are using the Transact Software and a commerce service provider to host the solution. For detailed information about each of these versions, access www.openmarket.com and review the ShopSite features section of the Web page.

Folio supports the efforts of the information publisher. The software offers the functionality to manage and distribute content from your intranet to the Internet. The Folio product incorporates authoring and content management, production, distribution, and commerce components. The authoring and content management component allows multiple participants to easily collaborate on a document from start to finish. Features such as document version control, content approval, and graphics and page layout are provided. The production component, called LivePublish, integrates with existing authoring tools and content management systems. The software provides the ability to take individual finished pieces of content from various standard file formats, to organize the information, and then to index the complete content for full-text searching. The distribution feature, also handled by LivePublish, automates the delivery of multi-gigabyte collections of information. To perform this task, it provides searching capabilities, integrated table of contents, custom query templates, ranked hit lists, and dynamic reference windows to help users find information quickly. In order to publish the content on the Web and receive payment for the material, OpenMarket offers a product called SecurePublish which tracks information usage and controls access based on the negotiated license. This information is then delivered to both the corporate knowledge officer and to the publisher. OpenMarket's Transact (discussed below) is used to sell the content on the Internet by supporting payment transaction processing, customer service functions, and account maintenance.

In addition to its commercial Internet presence, Open-Market offers a solution, called Transact, to support those companies whose business processes feature order management. When deploying Transact, you can integrate it into a third-party commerce package or into OpenMarket's commerce software solutions.

Transact allows merchants to manage orders from the initial purchase of an item to the updating of internal inventory systems. Transact provides capabilities to capture, store, and integrate with existing systems, and to process purchase orders, checks, and other forms of payment. In addition, the software allows you to fulfill orders and send that information immediately to your customer service department. Other features include online customer authentication and authorization, automated tax and shipping calculations, and online customer service. Transact even supports partial shipments and backorders.

Advertising and marketing capabilities are provided through tools that enable sales analysis, buyer behavior, and buyer profiles. There are also features to provide advertisements, special offers, and other promotions on the commerce site. Digital offers and digital coupons are integrated into the solution by employing e-mail and push technologies. This technology distributes special offers and discounts to targeted customers and businesses.

Transact supports purchase orders, credit cards, procurement cards, microtransactions, and SET. The system can also support debit cards, smart cards, frequent buyer points, and other payment types. Support for automatic subscription renewals and partial billing and credits with payment reauthorization is incorporated in the tool.

Integration Tools

OpenMarket supports a wide variety of integration tools. The underlying Web server used by Transact is OpenMarket Secure Web Server. OpenMarket licenses Netscape's Enterprise Server and fully integrates it into the Transact software. Oracle 7 and Sybase are the default relational databases used for storage and management of the commerce data. ERP systems, such as SAP, Baan, and People-Soft, may also be integrated into the solution. OpenMarket provides API for order fulfillment and other components of its architecture. Additional tools also provide integration into existing EDI systems.

OpenMarket and Third-Party Plug-ins

Several companies have partnered with OpenMarket to provide integrated solutions. The companies include Fast Engines, Inc., which offers open cross-platform performance middleware for Web-enabled applications, First Data Corp, Paymentech, CyberCash, and VeriFone, which offers customers the ability to securely process end-to-end credit card payments over the Internet. Tax computation for OpenMarket relies on a software module from TaxWare International, which is included in Transact. Other vendors are also integrating their solutions. For updated plug-ins, visit the OpenMarket Web site at www.openmarket.com.

Commerce Hosting Service

OpenMarket encourages commerce hosting partners to host their solutions. Most of OpenMarket resellers provide hosting services for small, medium, and large corporations. There are hosting services for both their commerce suite and Transact software. Commerce hosting partners include ADN (www.adnc.com) and AT&T (www.securebuy.com). OpenMarket has a list at their site of all of the commerce hosting providers from around the world. OpenMarket has provided one of the best options for those companies that want to use their software solution, but want to avoid the high cost and resources required to manage their own e-commerce site.

Customers

Companies that deploy OpenMarket solutions represent a global "Who's Who" of business. SONY, one of the world's leading manufacturers of audio, video, television, and information technology products, is using OpenMarket's Transact to power its VMO Direct store. Now it is easy to purchase award-winning Sony VAIO notebooks and desktops, Trinitron computer displays, Digital Mavica cameras, accessories and peripherals. Sony refers to its VMO Direct site as "the way online shopping should work."

Acer, the third largest PC manufacturer in the world, uses OpenMarket Transact to power its online store. With more than 28,000 employees staffing 120 enterprises in 44 countries, Acer produces a wide variety of computer products, including motherboards, DRAM, CD-ROM and CD-RW drives, ASICs, BIOs, memory products, and other primary components.

Based in Seattle, RealNetworks develops and markets software products and services designed to enable users of personal computers and other digital devices to send and receive real-time media using today's infrastructure. RealNetworks is using OpenMarket's Transact to power its online Real Store.

Tandy Corporation's Radio Shack is the nation's largest consumer electronics chain and one of the most trusted electronics retailers in the United States. With more than 7,000 stores and dealers, RadioShack sells more wireless telephones, telecommunications products, and electronic parts and accessories than any other retailer. Now Tandy Corp. is making shopping with RadioShack easier than ever before. OpenMarket's LiveCommerce makes it easy to find just what you are looking for in the Sprint Store at RadioShack.

C & K Components, Inc. is a worldwide leader in the design and manufacture of switches and other electromechanical components. C & K's e-commerce Build a Switch system is an example of using OpenMarket's LiveCommerce to increase customer satisfaction while reducing costs.

Milacron Inc., a leading manufacturer of tools and systems for processing plastics and metals, offers more than 50,000 items for sale at the site. Milacron is leveraging the Internet to change the way they sell metalworking supplies to more than 100,000 small job shops, significantly expanding their ability to serve this market.

AT&T, one of the world's largest providers of communications services to businesses, is using OpenMarket software as a key component of its new SecureBuy service. AT&T SecureBuy transaction-enables Web sites for companies wishing to conduct Internet commerce. In this capacity, AT&T enables businesses to outsource the complexity of Internet transaction processing.

Barclays Bank, has developed BarclaySquare, a leading online shopping mall in the United Kingdom. Barclays Bank uses OpenMarket Transact because it allows merchants to offer time-based discounts, coupons, and joint-incentive programs, enabling Barclays' merchants to draw more buyers into the mall and keep them there longer.

INTERSHOP

The final vendor profiled in this chapter, INTERSHOP, supplies e-commerce solutions for merchants, enterprises, and commerce hosting providers. Their Merchant Edition incorporates storefront capabilities, order processing, inventory management, and marketing and promotion functionality. The Enterprise Edition solution provides more sophisticated features for system integration, and includes consulting services to produce custom integrated solutions.

Several companies use INTERSHOP software products. These companies (and solutions) include Bosch for order management, Celestial Seasonings, and Electronic Arts for business-to-consumer catalog applications, Hewlett Packard for generating enterprise business-to-business orders and communicating information to reseller channels in local languages and currencies, and Mercedes Benz for online catalogs.

Features and Management Tools

INTERSHOP's solution for merchants is aimed at providing basic storefront functionality to sell products and services. INTERSHOP Merchant Edition contains basic storefront and catalog features that allow you to assign multiple attributes to each product. This product includes several built-in features for inventory management. For example, there is a Purchasing Manager that informs the merchant when inventory quantities have reach minimum levels and alerts you to generate a purchase order sent directly to a vendor. A Customer Manager component records customer and supplier information, such as number of visits, purchases, and the type of business.

The Store Manager component provides information on order processing. This component tracks the entire order process from the time of purchase to the product leaving the warehouse. The software also generates invoices and packing slips, and records payments. Store Manager also tracks debits and credits, whether invoices are fully or partially processed, and when orders are cancelled.

Statistics & Preferences Manager is used to create discounts, to define shipping methods and costs, to generate reports on store traffic, customers, and product turnover, and to specify tax rates.

The advertising and marketing components include features for cross-selling, discounting, placing banners, and creating customer profiles. The Profiles component creates specific pages and content based on the products customers select and the information they provide while at the site. When the customer returns, these preferences display.

Integration Tools

INTERSHOP includes Sybase's Adaptive Server XI to store customer and product information. INTERSHOP also includes a set of cartridges (their terminology) used to integrate with back-end systems as well as some third-party plug-ins. SAP's R/3 order and product management systems can also be integrated into the INTERSHOP solution.

INTERSHOP and Third-Party Plug-Ins

Some of INTERSHOP's tools include Language Pack for Spanish, Search Accelerator Cartridge, which provides customers with global search capabilities throughout all stores within an INTERSHOP site, and a cartridge for electronic software distribution. INTERSHOP software also provides seamless integration with CyberSource, CyberCash, Veri-Fone and WorldPay CurrencyPay payment systems for credit card and other payment transactions. A cartridge integrates directly into OpenMarket Transact 4. INTERSHOP storefront features integrate with OpenMarket's Transact 4 order processing functions. There are also several other ISVs that provide plug-ins to the merchant component. To get a listing, go to INTERSHOP's site at www.intershop.com.

Commerce Hosting Services

ePages, a hosting edition cartridge offered by INTERSHOP, allows service providers to sell inexpensive, entry-level commerce stores to a mass audience. This edition suits businesses that want to host e-commerce sites for other companies using INTERSHOP's commerce technology. INTERSHOP also caters to a number of hosting providers within the US and internationally. A complete list of hosting providers is available at INTERSHOP's Web site, searchable by state or country.

INTERSHOP's solution for enterprises addresses the requirements of larger companies. INTERSHOP Enterprise Edition includes all of the components available in Merchant Edition. This edition also integrates with Sybase's Adaptive Server XI database or Oracle 8 as the back-end systems to store and manage customer and product data. A Developer Customization License allows professional solution providers to modify and enhance the database. Developer documentation describes template language extensions, server-side scripting, extension interfaces, payment APIs, hooks, and shared libraries. This edition also includes INTERSHOP Search Accelerator Cartridge and the INTERSHOP Developer Kit. Available, but not included with the software, are cartridges for SAP, EDI, and mainframe systems.

Customers

A variety of customers have successfully deployed INTER-SHOP solutions. Bosch streamlined order management by integrating with heterogeneous server systems. The world-renowned tea maker, Celestial Seasonings, uses a self-hosted e-commerce application for selling their products online, including a new line of herbal supplements. The world's largest independent vendor of entertainment software, Electronic Arts (EA), uses INTERSHOP to sell popular products to computer game enthusiasts. Their online shop connects to an Oracle back-end and perfectly matches EA's decentralized business structure. Hewlett Packard and INTERSHOP developed a revolutionary concept to generate orders and communicate relevant information to reseller channels in local languages and currencies. The Mercedes-Benz online store, featuring integration of the UPS shipping module, is a digital extension of the boutique at the Visitor Center in Tuscaloosa, Alabama. Performance Snowboarding hosts an online store catering to a young, active, and upwardly mobile customer group, with features to match an equally progressive business model.

This chapter has surveyed some of the top e-commerce software vendor solutions on the market, including Microsoft, IBM, BroadVision, OpenMarket, and INTER-SHOP. These solutions, especially with vendor and third-party plug-ins, satisfy most business needs. However, before you select a vendor solution, determine how your company fits into the e-commerce marketplace. That is, define your vision statement, goals, strategies, and any refined business processes. If you understand fully how you intend to exploit this compelling opportunity, choosing a solution becomes painless and profitable.

Reading through the chapter, you may have noticed that some vendors provide solutions that cater to selling to other businesses versus selling to consumers. Other software solutions include features that support specific certain industries, such as banking, manufacturing, and travel. Still other software vendors provide solutions geared toward selling digital content versus selling products that are physically shipped to customers.

Once you know where to move your business in the electronic commerce marketplace, judge each vendor based on the five basic guidelines outlined in the chapter, including dominant features, commerce server management tools, integration tools, vendor and third-party plug-ins, and hosting services. Then just choose and deploy.

8

Payment Systems

- Options available for online purchases
- Providing credit card and micropayment methods
- Ensuring payment security
- Establishing customer trust

C ommerce is defined as buyers and sellers agreeing upon what is being sold and at what price. The seller delivers goods in exchange for payment. Separated into discrete parts, a transaction is comprised of an offer, authentication, payment, and delivery. In the virtual world, screens display images of goods and reflect the buyer's acceptance of a purchase. FedEx or FTP typically delivers the hard goods or digital content. The exchange of offers and delivery of documents via formal protocols has drawn limited market acceptance, primarily at business-to-business sites. Nevertheless, the secure exchange of the buyer's payment has been the first technical and psychological obstacle for most electronic commerce efforts.

An e-commerce site and associated systems must protect the interests of both the buyer and the seller by providing security and integrity. Security protects the seller from illicit use of the payment instruments, while shielding the buyer from misrepresentation. Integrity guards the buyer from an unauthorized disclosure of the transaction and shields the seller from disputes about the timing or terms of a sale. These requirements significantly raise the bar of complexity for Web sites and protocols.

Electronic payments generally are representative of real-world exchanges. Payment protocols provide an electronic counterpart to paper-based methods of exchanging goods. In fact, if performed correctly, electronic purchases prove far more secure than other purchases, as traditional, paper-based methods frequently display complete and unencrypted credit card numbers.

Most credit card transactions within the United States economy occur over dial-up lines between merchants and cardholders. In the Internet economy, many sites still rely on these protocols; this is especially true for merchants with existing retail systems in which a Web site acts as a counterpart to brick-and-mortar storefronts.

Electronic payment systems handle the monetary exchange transactions for goods and services. Similar to traditional monetary instruments such as bank notes, drafts, credit cards, etc., the e-commerce world employs a variety of tools. However, each tool can be judged on its conformance to the following requirements: independence, security, privacy, anonymity, transferability, divisibility, ease of use, and total cost.

- What are the various categories of electronic payment systems?
- What category dominates electronic commerce?
- How do transactions work in the physical world?
- How do transactions work online?
- What components need to be in place to enable electronic payments?
- How easy is the transaction for the parties involved, including the consumer, the merchant, and the financial institution?
- What are the actual costs of a transaction?
- How do I set up credit card verification?
- What protocols ensure online security?
- How do my customers develop a sense of trust?

Payment System Categories

Electronic payment systems fall into a variety of categories, based on their function. Micropayments, although measured in pennies, will enhance several aspects of e-commerce when a more innocuous business model is implemented. Smart cards, either disposable or reloadable, store monetary value on a card that can be debited. Electronic billing enables the presentment, payment, and posting of bills on the Internet. Open exchange relies on traditional bank or credit card clearing procedures to process a transaction, typically initiated by an e-mail message or HTTP form. Secured linkage is the encrypted transmission between merchants and customers where security-enhanced software and front-end browsers provide authentication. A trusted third party, as a mediator, handles all credit or check clearing functions between merchants and customers. After the process is complete, the trusted party provides the transfer of funds. Digital check, another category of electronic payment systems, replaces the traditional check with the security of an encrypted transmission. Digital cash resembles cash with its serial numbers, but it travels electronically.

These categories represent generalizations about the function of electronic payment systems. As a practical matter, your customers are likely to compensate you for goods or services in one of four ways: micropayment, smart card, electronic billing, or credit card.

Micropayments

Imagine you want to read an article from a subscription magazine published on the Web. Instead of paying for a year of subscription service or even a single issue, you might be willing to pay for only the specific article. Micropayment technology allows consumers to pay for digital content in increments of cents or fractions of cents. Such content typically includes music, videos, published articles, or software. The advantage of micropayments is that funds do not have to be reloaded for each transaction and minimum transaction amounts are much lower than for credit cards.

A typical micropayment transaction consists of four steps. First, the customer sends a request to load funds using his or her wallet software. Then, via the banking network, the funds are verified as available in the customer's bank or credit card account. Bank accounts and credit cards are handled differently as sources of funds. After approval, credit card funds are available immediately, while bank account funds might require up to a five day wait. After the money is transferred into the wallet, the customer can make micropayments at Web-based stores. The wallet automatically debits the amount of purchases from the balance. One of the main advantages of the micropayment method is that funds do not have to be transferred from a credit card or bank account each time a purchase is made.

The challenges of reasonably implementing micropayment systems are two-fold: delivery of the technology and charging for digital content. Several companies have established reliable business models based on micropayment technology, particularly on the delivery side. Qpass sells content from publishers such as The Wall Street Journal Interactive Edition on a short-term or per-article basis. Cybergold allows users to purchase digital content such as MP3 songs, software, and video files. Both companies have set up micropayment programs to address the challenges of charging small amounts online. These companies receive a portion of their revenue as fees from content partners.

One disadvantage to the micropayment system is apparent from the customer's perspective. Typically a customer might not object to paying five or ten cents to read an article or to access some specific information. However, multiply this experience twenty times within a single Web session, and your customer becomes exasperated. The merchant, in this example a publisher, would increase his revenue stream by charging a per issue, quarterly, or yearly subscription.

Smart Cards

People use many terms to describe smart cards: chip card, integrated circuit card, personal computer in your wallet, or database on a card. The smart card is an integrated circuit chip embedded in a small piece of plastic. Some models of these chips can hold 100 times the information contained on a standard magnetic-stripe card. The chips also adds "smarts" to the card, allowing it to store and to process information. Smart cards are available in disposable (or memory) and reloadable (or processor) versions. Disposable cards have value the user can spend. In contrast, most reloadable smart cards have more memory, more information, and a higher level of security. They can contain multiple applications on a single chip, manage several passwords, and use authentication and encryption techniques to combine freedom of use with security.

Around the world, there are more than 100 smart card programs in operation. The most common use of smart cards are telephone cards. A major card manufacturer estimates that 420 million phone cards were sold worldwide in 1996. Many leading banks also issue smart cards that function as credit and debit cards. In the United States, as part of its Smart Ship Project, the Department of the Navy uses stored value cards to replace all cash transactions aboard the Aegis Class Cruiser, USS Yorktown. In Mexico, Bancomer, S.A. issues stored value smart cards to truck drivers for automated toll collection in remote parts of the country, eliminating most error, theft, and abuse in reporting expenditures back to the company. In Thailand, Thai Farmer's Bank issues smart cards that combine a customer's emergency medical records with smart bank IDs and electronic purse features. Eventually, smart card technology may replace cash, checks, and credit cards in transactions that require safety, application logic, and convenience.

Smart cards offer a variety of benefits to consumers. As a replacement for cash, smart cards increase security and simplicity, and reduce paperwork and record keeping efforts. You no longer need to carry as much cash, you can print a transaction history at a convenient ATM, and you can avoid time-consuming forms. As a replacement for coins, smart cards increase transaction speed and convenience, and reduce bulk. You no longer need to load multiple coins into a vending machine or carry around coins for routine transactions that require exact change.

Smart cards also offer a variety of benefits to merchants, financial institutions, and other card issuers. As a merchant, you will appreciate higher sales, faster transactions, easier bookkeeping, reduced costs, and fewer losses. Smart card transactions are typically larger and faster than comparable cash transactions. In addition, electronic payments are easier and less expensive to handle and to reconcile than cash payments. As a card issuer, you gain more customers, lower overhead, and less fraud through smart cards.

VeriFone, a division of Hewlett-Packard, offers a range of point-of-sale (POS) payment terminals and peripherals, integrated payment software solutions, and Internet commerce solutions. The company continues to explore new payment technologies, refine its product range, and actively promote industry initiatives, including U.S. smart cards and worldwide Internet commerce. For more information about smart cards, their benefits, and how to establish smart card systems, contact VeriFone at www.verifone.com.

Electronic Billing

Electronic billing lends itself well to both business-to-consumer and business-to-business efforts. Whether you are currently sending out paper-based bills or have not yet developed a billing system, consider the role that electronic billing might play for your organization and its e-commerce solution. One of the reasons for electronic billing's growing popularity is its clear demonstration of revenue generation. When electronic billing first started gaining momentum, billers and banks were looking at the technology as a way to reduce printing and mailing costs and to speed payment collection. Today electronic billing allows one-to-one marketing and dynamic information presentation. At its full potential, this method of billing allows companies to increase their contacts, deliver customized information, and provide links to other online services.

Electronic billing enables the presentment, payment, and posting of bills on the Internet. Presentment involves taking the static statement data, which is typically directed to printers, and hosting that information on an interactive Web-based bill presentment server. With the Web, merchants can customize the user interface to each individual customer. By processing payments online, electronic billing provides cost savings and cash management benefits to the biller that can consummate the collection process directly on their own Web site. After the statement is presented and payment is secured, the billing merchant must post to their accounts receivable system and update the account.

A typical electronic billing transaction is comprised of six steps. First, after verification (user name and password), the bill is presented to the customer. Then the customer reviews, analyzes, and pays the bill. The resulting transaction typically debits either a bank account or a credit card of the customer. The transaction is posted to the clearing house (ACH) flat file. A success or failure response of the verification and posting of the transaction information is sent back to the biller. Then an electronic customer receipt is generated, indicating success or failure of the transaction. Finally, the merchant's clearing house (ACH) file is reformatted and submitted for account settlement.

This burgeoning practice has attracted a variety of companies, each of which is a good source for more information. The large players include Microsoft and CheckFree, a company specializing in payment processing. Other electronic billing vendors include BlueGill Technologies, @Work, Novazen, and eDocs. Service bureaus and billing-service providers, including Pitney Bowes, IBS, and Bell & Howell, ease the transition to electronic billing.

Credit Cards

A credit card transaction is actually a loan made by a lending institution to a customer. If the customer pays back the loan by its due date, the loan is interest free. But if not, the rate of interest can be very steep. According to federal banking laws, the rate cannot exceed 25% annually. Consumer protection laws issued by the Federal Reserve Board also limit a customer's liability for the fraudulent use of their card to $50. This is one reason why banks sometimes wave this fee if the customer reports a lost or stolen card to them immediately. The report reduces the potential loss to the bank by allowing quick deactivation of the account. With this limit protecting the customer, credit card transactions prove to be the driving force for electronic commerce between businesses and consumers. While micropayments, smart cards, and online billing each play an interesting and important role in electronic commerce, the top three credit card companies handle over 98% of all purchases made on the Web. Contrast this figure with transactions outside of the Internet where only about 20% of all purchases are made with credit cards.

Most financial institutions, credit card verification systems, and e-commerce software programs integrate the credit cards of companies such as American Express, Carte Blanche, Diners Club, JCB, MasterCard, Novus/Discover, and Visa. In fact, MasterCard and Visa account for just over 70% of all purchases made by credit cards.

Credit card transactions are handled in much the same way, whether in the physical world or online. The components of a typical transaction include authorization and settlement. Authorization refers to the approval process of a credit card transaction. Settlement refers to the payment of various fees to companies that assist the transaction, as well as payment of the merchant by the bank.

Credit card authorizations can be separated into a series of finite steps. First, a consumer visits a Web site by using a standard Web browser. The consumer selects an item to purchase at the merchant's Web site and adds it to a shopping cart. After entering shipping and credit card information, the consumer is presented with a summary of the item, its price, and the billing information. The payment information is protected by Secure Socket Layer (SSL) encryption and forwarded with the order form to the merchant's commerce site. The server software adds the merchant's identification information to the packet. The secure payment request is forwarded over the Internet and is received through a secure firewall by a series of servers that pass it to the merchant's financial institution. From there, the packet travels to the customer's credit card bank to approve or decline payment authorization. The consumer's credit card bank sends its response back through the merchant's financial institution to the merchant's system. Typically, this authorization process takes 20 seconds or less to accomplish.

Figure 8.1 Credit Card Authorization

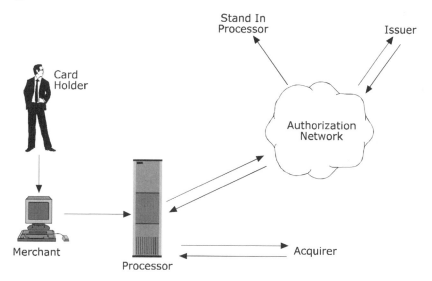

After the transaction is authorized, the consumer's credit card bank settles the transaction (a process also known as interchange). As a merchant, you should understand the process by which credit is settled in order to determine if accepting credit cards at your online store makes financial sense and if so, how you can structure your prices to account for these costs. The merchant delivers the goods to the consumer, and requests financial settlement. For a $100 transaction, the merchant receives $98 based on a 2% discount rate. That money is deposited in the merchant account at the merchant bank (also known as the Acquirer bank). The Acquirer bank receives $.67 of the $2 paid by the merchant (the cost of offering credit card service to consumers). The merchant bank pays the Association (Visa or Mastercard, for example) $.08. The issuer of the credit card (Issuer bank) receives $1.25 for owning the cardholder relationship. Eventually, the consumer is billed for the purchase, and either makes payment or incurs credit card debt. At this point, the entire transaction is complete.

Figure 8.2 Credit Card Settlement

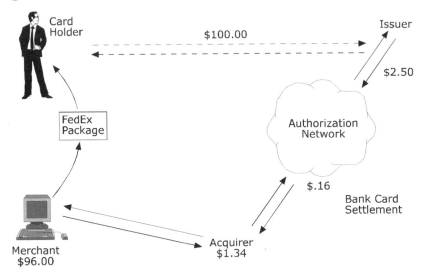

Enabling Electronic Payments

Enabling electronic payments over the Internet for your e-commerce business requires establishing an account with a financial institution and an Internet payment company, and also setting up your merchant register to accept payments. These payments originate from your customer who makes purchases by using an electronic wallet or by completing a form with purchase information incorporated into the merchant Web site. The following sections describe each of the components that must be enabled for electronic commerce.

Electronic Wallet

The software component that a consumer uses to make purchases electronically is referred to as an electronic wallet. The wallet allows the customer to store information on their desktop, eliminating the need to continually refresh information such as customer name, shipping and billing address, and payment information. Electronic wallets support several types of credit cards, digital cash, paper and digital checks, and purchase orders. Software versions of electronic wallets include vWallet (VeriFone), Commerce-POINT Wallet (IBM), and Microsoft Wallet (Microsoft). Most sites do not rely on wallets because their implementation forces the user to download software and to establish a relatively complicated configuration.

A popular alternative to the electronic wallet is a form that the merchant incorporates into the Web site. This form, filled out by the customer, captures contact information, payment information, and product choices. Most popular commerce solutions, usually via templates, allow these forms to be designed, developed, and deployed easily by the merchant. Although the resulting electronic commerce is form-based, instead of wallet-based, the remainder of the components perform the same functions.

Merchant Register

A merchant register represents the software a business uses to exchange payment and order information with cardholders and to collect payments from the cardholder's financial institution. Features of these registers include support for a variety of payment types, security protocols (SET and SSL), complete integration with a merchant's storefront technology, and financial institution support. A typical merchant register allows the merchant to accept payment via credit cards, micropayments, and smart cards, for both time-of-purchase and electronic billing. Most commerce solution vendors allow easy integration of CyberCash and VeriFone products into the register, eliminating the need for extensive programming. In addition, most merchant register software is supported by a wide variety of financial institutions. Versions of merchant registers include vPOS (VeriFone), CommercePOINT eTill (IBM), and CashRegister (Cybercash). Information regarding the products can be found at each vendor's Web site.

Internet Payment Service

An Internet payment service functions as an intermediary between merchant register software and the financial institution. Capabilities, provided by Internet payment services such as CyberCash or VeriFone, enable a merchant to accept online payments from customers. These payments are securely processed from a merchant's storefront through the existing system of financial institutions and credit card processing companies.

When considering various Internet payment services, focus on the key qualities of security, ease-of-use, and price performance. Ensure that the service you select provides security for credit card authentication, transaction confidentiality, server host integrity, and server virus protection. At a minimum, the service should provide Secure Socket Layer (SSL) protocol security and Secure Electronic Transaction (SET) encryption and authentication. An Internet payment service creates an easy-to-use experience by featuring real-time payments, scalability, easy installation and upgrades, and multiple payment methods. Real-time payments mean immediate bank approval of a sale, as well as expedient deposit and settlement of payments. Effective scalability allows smooth growth of transaction volume and additional payment features. Determine the complexity and workload required to install and upgrade software and hardware products. Your Internet payment service should allow your customers to change the methods that they use to pay for your products or services. Finally, the company should offer compelling price performance driven by its integration with your storefront and its relationship with a wide variety of financial institutions.

Financial Institution

A financial institution contracts with a merchant to enable the acceptance, processing, and deposit of credit card transactions. Merchants must maintain an account with a financial institution to receive credit for credit card sales. The financial institution deposits daily credit card sales into the merchant's account, minus any applicable fees, and delivers regular reports based on any transactions.

Acquiring financial institutions sometimes perform merchant services in-house and others outsource these functions to a third-party processor (i.e. First Data Corporation, Global Payment Systems, Nova Information Systems, or Vital Processing Services). A third-party processor is a company that provides credit card services to merchants on behalf of acquiring financial institutions. Services may include billing, customer support, reporting, authorization, and settlement services for merchants.

Establishing an account with a financial institution is reasonably uncomplicated. Merely contact the institution, describe your e-commerce configuration (what components you are using), and select an Internet payment service. Initially, make sure that the financial institution is authorized to underwrite credit card services to merchants. Direct the financial institution to set up the account to accept credit cards over the Internet using your choice of payment services. Determine the details of establishing a merchant account, including the application and approval process, the fees and charges, and the requirements for opening a deposit account for conducting business. Assume that the entire process will take 2 to 14 working days to complete.

Implementing Credit Card Verification

The final phase of enabling electronic transactions is to set up a credit card verification system. Each solution vendor has specific instructions for incorporating verification services into their commerce software. As an example, let us walk through the CyberCash procedure for implementing credit card verification. CyberCash has just released a service called CashRegister 3, a new Internet payment architecture that makes it easy to integrate storefronts, operate payment services, and upgrade to new services, standards, and options as they become available.

Implementing CyberCash technology begins in one of three different ways. You can use the services of a Merchant Development Partner to integrate and/or to host the Cyber-

Cash commerce-enabled Web site for you. You could implement the solution yourself by downloading the CyberCash software and integrating it into your storefront. Or rather than building a commerce-enabled Web site from scratch, select an e-commerce application or storefront solution in which CyberCash service is already integrated.

In order to accomplish the second option, perform the following steps. Collect and review information about Cyber-Cash and Internet payments (www.cybercash.com). Register with CyberCash and download the software. Apply for a merchant bank account. The bank accepts your application and then notifies CyberCash. CyberCash enables cash register functionality and e-mails you a notification. Configure your storefront for CyberCash. Then retrieve setup information from CyberCash and enter it into the storefront. Submit a test transaction. Assuming that everything works, ask to begin service. CyberCash notifies you that service can begin. Now you can open your store for business. More information about this specific implementation is available at www.cybercash.com.

A variety of companies provide credit card transaction solutions for merchants. Bank of America's Merchant Services division is involved in several Internet pilots to facilitate secure payments over the Internet. Charge Solutions provides Internet-based real-time credit card processing to merchants. Credit Card Network, a credit card authorization service, handles credit card transactions for merchants who have already set up their electronic storefront. Epoch Transaction Services offers secure, real-time billing solutions for membership and product-based Web sites. First Data, a huge provider of merchant processing services, handles credit card transactions in the retail, hospitality, supermarket, direct marketing, and health care industries. iBill provides credit card processing services. Paymentech offers full-service electronic payment solutions for merchants, third-party transaction processing, and total commercial card payment programs. Secure Trading, a British company, also offers credit card processing services for Internet electronic commerce sites.

Secure Electronic Transaction

One form of security, the Secure Electronic Transaction (SET) payment protocol is an Internet standard designed to provide a high level of security and anti-fraud assurances for payment card transactions over the Internet. Visa and MasterCard, along with technology partners such as Veri-Fone, developed SET with the goal of replicating the level of confidence that all parties to a transaction receive when an in-store transaction is conducted. The authentication process involves a series of security checks performed using digital certificates issued to customers, merchants, banks, and payment brands.

The four SET components include a cardholder wallet, a merchant server, a payment gateway, and a certificate authority. The cardholder wallet generates SET protocol messages that can be accepted by the three other components. The merchant server component processes payment card transactions and authorizations. The payment gateway component (run by an acquirer or a designated third party) processes merchant authorization and payment messages, and interfaces with private financial networks. The certificate authority component issues and verifies digital certificates as requested by the other components.

A transaction secured by a SET-enabled browser and a SET-enabled server follows these steps. The customer opens a bank account that features Mastercard or Visa services. The customer receives an electronic file (a digital certificate) that functions as a credit card for online purchases. Third-party merchants also receive digital certificates from the bank. The customer then places an order, either over the Web, via e-mail, or by telephone. The customer's browser receives and verifies the validity of the digital certificate from the merchant. The browser then sends encrypted purchase details. The merchant receives and verifies the validity of the digital certificate from the customer. The merchant then sends to the bank the order information, including the bank's public key, the customer's payment information, and the merchant's certificate. The bank then verifies the merchant certificate and the order information message. The bank digitally signs and sends the proper authorization to the merchant. Finally, the order is filled.

The SET protocol provides confidentiality, data integrity, verification of identity, and non-repudiation of disputed charges. The SET protocol uses advanced public key cryptography algorithms to assure that messages are transmitted and received confidentially—safe from the prying eyes of any uninvolved parties on the Internet. Cryptography algorithms generate the encryption (scrambling) and decryption (unscrambling) of information. The process of public key cryptography allows anyone to send an encrypted message to a designated recipient, using what is known as a public key. The recipient then uses a private key to decrypt the message. Therefore, only the designated recipient has the ability to read the message.

The SET protocol can also send credit card information directly from the consumer to the merchant's financial institution, without allowing the merchant access to the cardholder's account information. A consumer can therefore make a purchase with confidence that no fraudulent activity will take place with his or her credit card number after the card number has been transmitted. Merchants accepting SET transactions in this manner are protected from the liability of credit card numbers in their database, since they do not receive the credit card information.

When a message sent via the SET protocol is encrypted, a unique numerical value is generated for that message. If anyone attempts to tamper with the data contained in the message, the numerical value for the message would change and would therefore be detected. This process ensures the integrity of the message.

The SET protocol uses digital certificates to verify the identity of all parties involved in a transaction. When a consumer makes a purchase using the SET protocol, a digital certificate is sent from the consumer to the merchant and, simultaneously, from the merchant to the consumer.

The consumer's digital certificate functions as a signed credit card slip to the merchant. The digital certificate provides verification of the consumer's identity and offers a high level of assurance that the credit card is not being used fraudulently by the consumer.

The merchant's digital certificate provides the consumer with verification of the merchant's identity and leaves the consumer with a high level of confidence that the merchant is operating a legitimate business, and that the purchased items will be shipped.

The process of digital certificate distribution that takes place during SET transactions also holds the consumer and merchant accountable for information contained in the transaction. Therefore, if a consumer places an order for a product using SET and then receives the product, he or she cannot deny that the order has been placed. Similarly, merchants cannot later deny that they received the order.

Secure Sockets Layer

Another form of security, Secure Sockets Layer (SSL) is a commonly used program layer for transmitting payment information between a buyer and a seller. SSL, which was developed by Netscape, sends data back and forth between the cardholder's Web browser software (such as Netscape Communicator or Microsoft Explorer) and the merchant's Web server software (such as Netscape's Enterprise Server or Microsoft's Commerce Server), within a secure environment. SSL encapsulates data into transactions and guarantees the confidentiality and integrity of the individual blocks of the transaction.

The sockets part of the term refers to the sockets method of moving data back and forth between client and server in a network. A socket is defined as the endpoint in a connection. Passing sockets is a method for communicating between a client program and a server program in a network. Sockets are created and used with a set of programming requests (or function calls), sometimes called the sockets application programming interface (API).

The layer part of the term refers to the program layer between an application and the Internet's TCP/IP layer. The programming code for keeping your messages secure is contained in the program layer.

SSL uses the public key/private key encryption system from RSA, which also includes the use of a digital certificate. The secure environment for SSL is created through the use of public key cryptography, which consists of the encryption (scrambling) and decryption (unscrambling) of information. Public key cryptography allows anyone to send an encrypted message to a designated recipient, using what is known as a public key. The recipient then uses a private key to decrypt the message. As a result, only the designated recipient has the ability to read the message.

The SSL protocol is a powerful tool for the secure distribution of information, but does not address all of the risks associated with sending and accepting transactions over the Internet. For example, SSL establishes a secure session between a browser and a server. During the period when the browser is logged onto an SSL server, authentication between the browser and the server takes place. However, SSL does not authenticate the parties who are using that software. Thus, while cardholders using SSL can submit payment information free from the prying eyes of a third party, there is no way of verifying the identity of the online storefront that they are visiting.

Establishing Trust

Who are you? Whether you are a buyer or a seller, the initial requirement of any real-world transaction is identification. For example, when you, the shopper, walk into a franchise hardware store, let's say Home Depot, you are comfortable with the brands offered, the service you receive, and the store's return policy. You do not spend a lot of time wondering if you are going to be subject to theft or fraud. It is the merchant who may wonder who you are, and whether the company may be subject to theft or fraud. When you go online to shop, however, the tables turn. Now you the shopper wonder about the store. And the merchant has established layers of security to protect his interests prior to opening for business. From a merchant's perspective, establishing the trust of customers is critical to the success of an e-commerce solution.

The six primary components of e-commerce trust include seals of approval, brands, navigation, fulfillment, presentation, and technology. Seals of approval, represented by symbols like VeriSign and Visa, are designed to reassure visitors that sites have established effective security measures. Known as security brands, these seals of approval testify to the safety of a merchant's site, its technology, and the network behind it.

Brand refers to a company's implicit promise to deliver specific attributes, based on a company's reputation and a visitor's previous experience with its products. The concept of branding includes online and offline brand recognition, portal or marketing affiliations, community building, and the sense of a site's breadth of product offerings.

Navigation refers to the ease of finding what a visitor is seeking. Navigation is aided by understandable terms, the consistent placement of navigation elements, clear instructions to help shoppers make their way through a site, and simple, appropriate terms to describe site content.

How clearly a site indicates the way orders are to be processed, its return policy, and how well it explains the way customers can seek recourse to problems are all elements associated with fulfillment. The assurance that a customer's personal information will be kept secure and private is a key attribute of fulfillment.

Presentation implies the ways in which the look of a site communicates meaningful information. On the home page, a site's purpose must be clear to the first-time visitor.

Visitors evaluate technology largely in terms of speed and function. How well does a site's technology operate and how quickly does each page load? It is important to balance page loading speeds with helpful functionality.

How do you address each of these components to establish the trust of your customers? Add a policy statement explaining how transactions work and how returns are handled. Provide your corporate history. Add a variety of security seals, including VeriSign, Microsoft, and the Better

Business Bureau. Display a message indicating that any transactions that take place are secure and encrypted. And provide a toll-free number to allow users to pay by phone if they choose. When designing your commerce site, assume the perspective of your audience and address their needs.

9

Security

- Security threats
- Creating security policies and procedures
- Separating your network from the Internet
- Using encryption effectively
- Preventing viruses

C onnecting to the Internet creates a bevy of opportunities for a company, as well as many potential security risks. Often companies address these issues by focusing only on the hardware and software needed to keep hackers and viruses outside of their intranet system. Hardware and software considerations represent only one layer of security. Implementing solid security measures also means establishing policies and procedures that require users and network administrators to take precautions.

Security threats have been on the increase from both outside and inside corporate boundaries. Traditionally, 80% of security threats were generated from within a corporation. These threats typically included employees within a company gaining access to company systems through loopholes in security and by access to employee passwords. These percentages are changing due to the growth of the Internet and corporate extranets. Intruders from outside your corporate boundaries now represent a growing concern.

Research, conducted in early 1999 by the Computer Security Institute (CSI) with the participation of the San

Francisco Federal Bureau of Investigation (FBI) Computer Intrusion Squad, revealed that both internal and external security threats are on the rise. For those companies that acknowledged unauthorized use, 43% reported from one to five incidents originating outside the organization and 37% reported from one to five incidents originating inside the organization. This report was based on responses from 521 security practitioners in American corporations, government agencies, financial institutions and universities. The report offers some other striking considerations:

• Those reporting their Internet connection as a frequent point of attack rose for the third consecutive year; from 37% of respondents in 1996 to 57% in 1999.

• Unauthorized access by insiders also rose for the third consecutive year; 55% of reported incidents.

With security threats rising, you must ensure that you take a proactive role in reducing your exposure. However, it is important not to fall into "security paranoia" by implementing such high security measures that you prohibit your employees and customers from communicating effectively.

This chapter focuses on methods to reduce security risks in areas most vulnerable to hackers, viruses, and other forms of network intrusion. Specifically, we explore the following security issues: network threats, security policies and procedures, firewalls and proxy servers, encryption, digital signatures, authentication and access control, virus prevention, and auditing your network.

Keep in mind when developing a security strategy that you must balance the need for communication and information transfer between your company and customers, vendors, suppliers, and the public, with the potential for harm that this access creates. Understanding security risks, especially those associated with electronic transactions, helps your company design and architect a secure infrastructure.

Network Threats

Hackers continue to invent new techniques for accessing internal networks and personal computers. A Web site (www.alt2600.com) provides information on the latest sites attacked by hackers and links to tools that they use. The sections that follow provide information on some of the ways intruders compromise security on corporate networks and personal computers.

Port Entrance

Several hacker tools, available for free on the Internet, scan a system's open ports in an effort to gain entry. For example, Portscan 1.2 is a utility which allows the user to scan ports on any target system. The user specifies the target IP address. The program then scans all ports between port 1 and port 65536. The resulting information can be used to find loopholes in a security setup. In addition, the program identifies FTP and WWW services on any assigned port. A similar tool called IP-Prober allows open ports to be scanned in order to gain entrance to the network.

To prohibit intruders from entering open ports, disable all ports and only enable those that are required to gain access to the network. Once you enable the selected ports, use firewall security to prevent unauthorized access. The firewall section of this chapter describes port security in more detail.

Password Cracking

Password cracking represents another method hackers use to enter your network. Several cracking tools, available throughout the Internet, generate and test a series of potential letter and number combinations until they determine the correct password.

To prevent password cracking, follow these basic guidelines. Require users to create and use passwords of eight or more characters that combine alphanumeric elements. In addition, many system administrators regularly run a tool called Crack. This tool tests users passwords to determine if they are easy to crack. This tool is available for download at www.cert.org.

E-Mail Spoofing

E-mail spoofing results when a user receives e-mail that appears to have originated from one person, but was actually sent by another person. The goal of this spoofing is to trick the user into divulging information or replying with information that is confidential.

One method of preventing spoofing is to use electronic signatures to exchange authenticated e-mail messages. Electronic signatures, also known as digital signatures, provide a mechanism to ensure that messages have not been altered during transmission and that the messages are from the person listed as the sender. E-mail offers options to send messages that incorporate digital signatures.

Virus Intrusion

Virus intrusion represents a constant threat. Virus exposure results from receiving e-mail, sending documents over the network, and even from installing software or copying files from another hard drive or diskette. In 1998, the Melissa virus, the CIH/Chernobyl virus and the Happy99 virus made headlines, causing huge damage to personal and corporate computers around the world.

On March 27, 1999, the Melissa virus began propagating via e-mail attachments. When attachments were opened in Word 97 or Word 2000 with macros enabled, the virus infected the Normal.dot template causing documents referencing that template to become corrupt. If the infected document was opened by another user, the document would propagate. Because the virus continued to propagate, mail servers experienced performance slowdowns.

On April 26, 1999, a virus called Chernobyl (CIH) started spreading. This virus infected executable files and then spread by executing an infected file. Eventually when the CIH virus activated, the virus began deleting information from hard drives and overwrote each system's BIOS.

On January 20, 1999, the Happy99 virus hit the Internet. The Happy99 virus is an executable file that displays fireworks and the words "Happy 99" on a user's screen. During its run cycle, the virus modifies files on the system.

With viruses attacking both the corporate and private world of personal computers, most computers feature virus protection software. If you do not have a version of this software, you can obtain it via the Internet from a variety of vendors. Remember to update the virus protection software to ensure you have protection against the most recent viruses. For corporate organizations, institute the appropriate policies and procedures to require users on the network to deploy and use the latest virus protection software.

Other Hacker Attacks

Hackers employ a variety of other methods to attack servers around the world. This section describes a few more areas of risk. Most of these hacking methods are preventable using sound security policies and appropriate software and hardware implementations.

The "ping" command can be used to exploit your system. Several free hacking tools allow hackers to send either oversized packets or continuous packets to your server. The result is a system interruption, or even a crash. For example, Ping-g is a tool used to specify an IP/Host address to continuously send a specified amount of ping packages. Another tool is the Ping of Death, used to send oversized packets to the server, resulting in a system crash.

Most operating systems now include the functionality of accommodating these larger packet sizes. In addition, if a hacker is using a tool such as Ping-g, monitor your system activity. If necessary, deter the sending system from accessing your server.

A tool called SScan performs probes against servers to identify services such as Telnet and FTP. You may be vulnerable to access by an intruder. The tool can be configured to automatically execute scripts of commands on your system, causing major problems. To prevent the SScan tool from locating vulnerabilities, use firewall filters to monitor any TCP/IP application ports and to detect unauthorized scripts running on the system.

Security Policies and Procedures

Any concerted effort to counter common network threats begins with comprehensive administrative policies and procedures. User-level security focuses on ensuring that employees are aware of the importance of security within the organization. Administrator-level security ensures that hardware, software, and user access is documented and granted according to the established guidelines. The following section describes the most critical areas to consider when defining security policies and procedures. Depending on your company's needs, some requirements may vary in how intensely they are monitored. For example, in environments of highly confidential information, more resources (time and money) may be expended to deploy multiple layers of firewall security (instead of just a single layer).

User-based Security

User education and security awareness programs should be the starting point for developing a secure environment with the support of your employees. Often in companies, employees work unaware of the implications associated with the following situations.

Do you allow others to access the network with your user name and password? You may have asked, "Can I log on with your name and password to get access to the network." The response is typically, "Sure, no problem." Actually, the problem has just started. Ensure that personnel within the company understand the importance of not allowing others to use their login name and password. If someone is logged on as another user, it becomes difficult to determine who actually caused a security breach.

Do you leave your computer on when you leave your work area? Assume that a spy is just around the corner. This situation creates a risk that most people may not even consider. It is time for lunch or a meeting and you are working on a document. Instead of logging off the system and turning off the computer, you just walk away from it. A virus can be deployed within 30 seconds or less. Again, establish a policy that requires users to log off prior to leaving the work area. This simple action can save a company a lot of time and money.

Do you write your password on a note lying next to the computer or stored in a drawer? Our workplace has become the "login name and password" game. Each system requires multiple login names and passwords. However, writing them down on a piece of paper means not having security at all. Ensure that users do not practice the "sticky note" habit. Remind them how critical these user names and passwords are to protect their data as well as company information.

Users must understand that the information they are granted access to is secure and is confidential. Users must be informed that they are given different degrees of access to sensitive company information based solely on their user name and password.

Companies can take additional steps to help users manage confidential information. For example, when sending e-mail or documents electronically, an option is provided to include the word "Confidential" in red letters across the top. This tagging informs employees who may have access to this information that the document is of a sensitive nature and should not be sent or discussed with others in the company.

Administrator-based Security

Administrator security policies must be established to protect company information and computer systems. These rules dictate both a protective arrangement of hardware and software and a series of regular practices to ensure an organization's technical health. Setting security policies and procedures begins on a physical level and includes operating system security, Internet security, remote access rules, virus prevention, and commerce application protection.

Physical server security is the first level of policies and procedures that must be established. Keep servers in a locked room with controlled access. If someone can get to the server, they can often very easily access the information. Make sure these servers are secure and register each person that enters and exits the server room.

Protect keyboard access with passwords. If controlled access is breached by an intruder, the security afforded by passwords may protect your company. Exercise reasonable caution by ensuring password lengths are more than eight characters and by using alphanumeric characters. Systems requiring double passwords to access information might be appropriate in some circumstances.

Establish security procedures to prevent unauthorized users from gaining access to the network. Some companies require badges and digital cards to access computer rooms. Other companies require passwords prior to allowing access to the computer center. However, based on our corporate experience, administrators have often given their access information to outsiders, such as consultants working on site. Establish and enforce policies and procedures for these situations. You do not know who may be lurking in the hall.

The operating system requires its own set of security rules and regulations. Even if you have set security for the Web server through firewalls and proxy servers, access through the operating system remains a big risk. The following sections suggest a variety of security policies.

Define user, server, and group access rights to files and directories. You must determine who has read, write, delete, and modify access to directories and files on the system. It is also a great idea to setup a "no access" group. This is especially useful if someone from the company leaves the organization. You can easily and immediately place that user in this group, preventing access to system files.

Require passwords to be greater than eight characters in length. This will make it more difficult for users trying to access the system. Furthermore, require a combination of alphanumeric characters which are more difficult to crack.

Another tip is to set up the system so users are required to change passwords every two to three months. Again, this provides another layer of protection.

Place system files on a separate partition. This will make it easier to manage the information. For example, place all of the system files on the C: drive and give only administrators access to the information. Once you have done this, log all additions, deletions, and modifications to these systems files. These are the files hackers seek. Also monitor server log files for suspicious queries or unusual system access.

Limit almost all root access. If someone has root access, they can perform basically any command on the system. This includes adding, deleting, and monitoring of files.

Internet security policies and procedures include defining Internet access and ensuring port security. Define Internet (external) access for buyers, suppliers, trading partners, and customers as a whole. The best method to set these policies is to determine the main groups that require access and then determine their level of access to the information on your site. For example, you may have company A, company B, and anonymous Internet customers. Once you have set up these groups, setting up site access proves much more manageable. Imagine that company A has access to certain products you are selling at the site. When members of company A log on, they immediately are placed at the site which contains products and services specific to their company. In addition, you can provide specific vendor discounts and server access (based on user login name and password).

Port security is also critical to Internet security policies. Internet port and security configuration allows enabling or disabling port 80 for HTTP access and enabling and disabling port 443 for SSL access. When Web clients access the server, they access port 80 by default, similar to a door that allows access to the Web server. When this door is opened (enabled), they have access to the information at your store. This port should be enabled to allow customers to access the information at the site. However, there are also several other ports that provide an open door to other applications on the network. For example, 20/21 is used to access the FTP server on the network. This port should not be displayed if you do not plan on having an FTP server for your e-commerce site. The bottom line is to only allow access to those applications, such as HTTP, that are required by your customers on the Internet.

Firewall and proxy server security adds another layer of security to your site. The firewall is used to protect your intranet or internal network from the Internet. As a complement, proxy servers limit the sites which your internal users can access on the Internet. We discuss firewall and proxy server security in more detail later in the chapter.

Policies and procedures are also required when providing access to servers that have dial-in capabilities. When setting up a remote server, ensure the server is behind the firewall and follow strict access and operating system security procedures. Other forms of security include using dial-back and password protection for modems. Keep in mind that even if someone is dialing into the network, as soon as they are online, they are participating in your environment as another node on the network.

Policies and procedures on virus protection software must also be established for the company. For example, require that virus scanning software is installed on all computers and that updates are provided on a regular basis. Since new and updated viruses are a constant threat, a systematic approach to distributing these updates to users within the company helps protect the systems. Other measures include prohibiting users from copying/installing files or programs from home onto corporate computers. We discuss viruses in more detail later in the chapter.

Commerce application security includes defining those users and groups that have permission to perform specific tasks on the commerce server. Ensure that there are defined policies about who has access to perform particular tasks on the commerce site. Most commerce server software packages provide the capability to create user groups and assign access rights to each level of your storefront. These rights define who has the ability to add, modify, or delete product information, approve customer orders, modify site design, and they include administrative rights to perform all tasks on the Web site.

Separating Your Net from the Net

Firewalls and proxy servers provide security between your intranet and extranet and/or the Internet. Organizations install firewall software to protect their intranet from unauthorized external access. A firewall works at the hardware or software level to control access to the internal system. Firewalls implement such security measures as packet filtering, application-level gateways, circuit gateways and are often used in conjunction with a proxy server. A proxy server mediates between client requests in your company, such as HTTP requests from a browser and servers outside your firewall. Proxies mask the return address of the requesting computer, providing secure anonymity for users and denying any potential targets. Some proxies run virus detection programs on incoming packets. The following sections describe firewalls and proxy servers.

Firewall

In simple terms, a firewall consists of a server and two network interface cards. The external network card communicates with the Internet or extranet and the internal network card communicates with the intranet or internal network. In Figure 9.1, the firewall is placed between the internal servers and the external servers. This firewall design provides protection against external users accessing the internal network.

Figure 9.1 Firewall

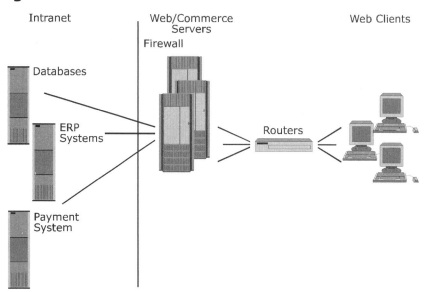

Firewalls provide a method to filter information that is transmitted over the network (sometimes called a datagram or packet). A packet-filtering firewall examines where the packet is going and what type of information it contains. It then checks whether the network allows delivery to that destination and allows that type of information to enter or exit your network. If the packet passes these tests, it is allowed to continue to its destination. Typically, packet filtering is implemented with router software.

Packet filtering provides a way to filter IP addresses based on IP addresses and ports. For example you could allow only known IP addresses access to the network. On the other side, you can deny access to all unknown or unreliable IP addresses (including competitors).

Another example of firewall security involves denying all outsiders access to port 80, resulting in no one from outside the organization accessing the HTTP server. If you decide not to implement this type of security, consider an alternative. The same result could be achieved by assigning another port to the HTTP server. Those intruders who attempt to violate the system security via port 80 would be unsuccessful.

What firewalls *can* do for your network:

- Prevent unauthorized logins
- Block outside traffic from the inside of your network
- Filter unwanted network traffic
- Provide security and access logs
- Act as an effective tracing or monitoring tool
- Enforce security policies

What firewalls *cannot* do for your network:

- Protect against traffic that does not flow through the firewall
- Protect against bad, unrealistic, or nonexistent security policies
- Protect against viruses
- Protect you from data-driven attacks (that is, e-mail, FTP or copied data, executed on the host)
- Protect you from data theft by authorized users, such as industrial spies or disgruntled employees

When planning your firewall architecture, the first step is to define firewall policies. The planning process is often the most time-consuming and cost-intensive. At this point, you must decide who has access, what servers they have access to, and what elements to monitor on the system. Based on security needs within the company, select the vendor software solution that addresses the security requirements defined for your corporation's architecture. Table 9.1 lists a variety of major firewall vendors, their products, and contact information.

Table 9.1 Top Firewall Vendors

Company Name	Product Name	Web Site Address
Ascend	Ascend Secure Access Firewall	www.ascend.com
Cisco	Cisco IOS Firewall	www.cisco.com
Sterling Commerce	CONNECT Firewall	www.sterling.com
CyberGuard Corp.	CyberGuard	www.cybg.com
LanOptics	Guardian	www.lanoptics.com
Microsoft	MS Proxy Server 2.0	www.microsoft.com
Check Point Software Technologies	Fire Wall	www.checkpoint.com

Proxy Servers

A proxy server functions as an intermediary between a secure network and a non-secure network. Most firewall products offer proxy server capabilities. With a proxy server, you can monitor information such as who has access to particular sites on the Internet. If necessary, you can prevent users within the company from accessing specific servers on the Internet. For example, the user can access a site if the Web server name is www.microsoft.com. However, if the user then tries to access www.playboy.com, the proxy server has been configured not to allow that access. This situation is illustrated in Figure 9.2.

Proxy servers also include a caching service for Web pages. The next time a user accesses a previously cached paged, response time improves because the server does not have to access the Internet to reproduce the page. The process of caching often dramatically improves response times. A proxy server receives a request for an Internet service, such as a Web page request from a user. The proxy server passes its IP filtering requirements (firewall) and looks in its local cache of previously downloaded Web pages (cache services). If it finds the page, it returns it to the user without needing to forward the request to the Internet. If the page is not in the local cache, the proxy server uses one of its own IP addresses to request the page from the server out on the Internet. After the page is received and filtered, the proxy server forwards the page to the user.

Figure 9.2 Proxy Server

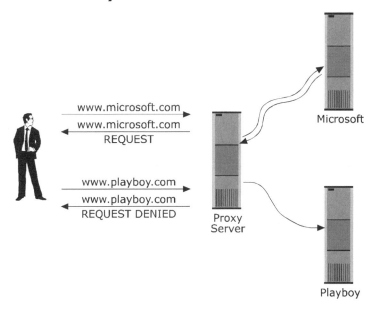

When setting up a proxy server, configure it as application-based or generic. Depending on how you plan to monitor user access to the network, you can select either of these two options.

An application proxy uses the same method as a packet filter. It examines where the packet is going and the type of information it contains. However, an application proxy does not let the packet continue to its destination. The application proxy delivers the packet for you.

An application proxy firewall is a server program that understands the type of information that is being transmitted, for example, HTTP or FTP. Application proxies control the flow of information between internal and external clients and servers. This option may be used if you want to control access to FTP sites, but not specific Web servers.

Web browsers have built-in settings to specify proxy server configuration. For example, the browser contains a setting to require a proxy prior to accessing the Internet, FTP, Gopher, or Web servers.

A generic application proxy is similar to an application proxy, except that it does not need to understand the type of information that is being transmitted. For example, SOCKS servers can function as this type of proxy. This method may be used if you want to monitor traffic from your intranet to the Internet.

Firewall and proxy server planning and implementation are critical to securing your corporate information and helping to prevent unauthorized user access. Firewalls are used primarily to control traffic from the extranet and/or Internet to your corporate information systems. Proxy servers, on the other hand, are used to control sites which employees can access from the internal network to the Internet. In the following sections we discuss other areas which are required to prevent corporate information and data from falling prey to a hacker.

Encryption

Encryption is the transformation of data to a scrambled form, which can not be read or interpreted without the appropriate knowledge (translation) key. Encryption and decryption require the use of common information, such as a key or key algorithms, to scramble and unscramble data. Why is encryption/decryption so important? Cryptography is the only practical method of protecting information transmitted electronically. Contracts, documents, money orders, and other legal information travel across the Internet. With the threat of hackers, wiretaps, competitors and dishonest insiders, a method to scramble data while stored on the system or during electronic transmission is critical. Take a look at a few methods of data encryption in order to understand how to use this technology to empower your own business.

Symmetric Key Encryption

Symmetric (or single) key encryption features a single key to encrypt and decrypt information. This technology works well when sender and receiver can arrange ahead of time to exchange the key. In most business cases such exchanges are cumbersome, especially among multiple parties. Figure 9.3 illustrates symmetric key encryption.

Figure 9.3 Symmetric Key Encryption

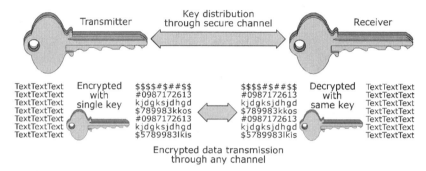

Asymmetric Key Encryption

Asymmetric (or public) key encryption requires two complementary keys to encrypt and decrypt information: a private and a public key. The public key, freely distributed, is only used for encryption and is not able to decrypt any encrypted documents or files. Only the complementary private key can decrypt the information. Figure 9.4 illustrates the process of asymmetric key encryption.

Figure 9.4 Asymmetric Key Encryption

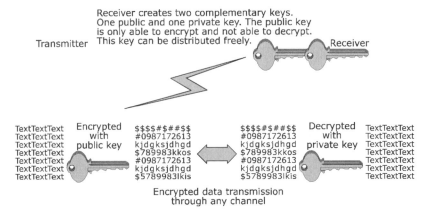

Encrypted data transmission
through any channel

Encryption and Decryption Algorithms

Two algorithms are widely used for the public-key encryption: the Rivest, Shamir and Adleman (RSA) algorithm and the Diffie and Hellman algorithm. The RSA algorithm randomly generates a very large prime number for a public key. Then the RSA algorithm relates the public key to another very large prime number, which functions as a private key. The integrity of the algorithm is based on the difficulty of finding the prime factors of a large integer. The Diffie and Hellman algorithm is based on the difficulty of computing discrete logarithms in a finite field generated by a large prime number.

Pretty Good Privacy

Pretty Good Privacy (PGP) freeware is a public-key encryption/decryption program, which can be downloaded from www.pgp.com. In addition, commercial versions are available. The program will automatically create public/private keys and key rings, and will distribute public keys through public servers on the Internet. It supports all major e-mail packages and allows encrypting and decrypting of data and files. PGP supports the Rivest, Shamir and Adleman (RSA algorithm) and the Diffie and Hellman algorithms.

Different software vendors offer one of these levels of encryption for sending e-mail or documents within an intranet or over an Internet or extranet. The benefit to encryption is data is not able to be read using a sniffer or other tool to capture the data as it passes over the communication link or while on the server.

Digital Signatures

The digital signature replaces the handwritten signature as certification of an individual's identity. Signatures provide information such as proof of origin, proof of submission, and proof of delivery. Digital signatures are used when sending electronic mail or documents on the Internet. These signatures authenticate the author/sender of a document. Authentication proves critical because e-mail messages and electronic documents are more susceptible to forgery (or misrepresentation) than other means of communication.

Authentication and Access Control

Login names and passwords are one of the most common methods of ensuring authentication. Authentication is used to identify users and programs and to grant access levels according to information on the network. Authentication also assures accountability. Since activity tracking is a standard tool featured in most corporate intranets and extranets, activity tracking such as file access, modification, and deletions can be set to track according to user logins. If some security breach does occur, a central point of contact remains (the person logged onto the network).

Authentication can be established by requiring a single login to access information on the network. This is the most common method of gaining access to resources, computers, and data on corporate systems. In more secure environments, double or even triple authentication methods are required for access. For example, in some environments, double authentication (requiring two individual logins) may be required to access extremely confidential information. With Domino software you can set multiple authentication requirements when adding new servers or users to an environment. Two or more people need to enter, for example, their user names and passwords prior to allowing these additional users into the environment.

Authentication technology may rely solely on logins and passwords. Login names and passwords are used primarily by employees to access network resources. More advanced methods are used to access sensitive data or servers. Smart tokens, smart cards, biometric measuring devices for thumb prints, palm prints, and voice, keystroke recognition, and even retinal scans can also prove effective. Based on the company's security needs, many variations can be exploited.

Viruses

A computer virus represents one of the most serious threats to the security and integrity of a computer system. Viruses damage systems by deleting files, modifying security, and changing hardware configurations, leaving some systems unusable. Computer viruses spread from user to user, program to program, and system to system. Due to this rapid growth pattern, a virus might be detected too late.

Warning Signs

Some viruses are obvious. Remember the Happy99 virus, for example, that displays fireworks on the screen? Others remain hidden. Basic symptoms of virus infection can serve as a cautionary warning. Although they are not always an indication of infection, these symptoms should raise your suspicion. Potential symptoms include:

- Time stamp or length of files changing unexpectedly
- Program load time increasing
- Unexplained writing to write-protected media
- Amount of available memory on a computer decreasing unexpectedly

- Executable files modified or deleted
- Workstation rebooting at random

How a Virus Works

A computer virus pursues two primary objectives for its creator. First, the virus copies itself into programs or files on a system. Second, the virus executes the instructions that the creator of the virus incorporated. This execution may be triggered by a date, time, or the execution of another program within the computer system.

The virus instructions may include commands to display messages on the user's screen, format disks, oddly reboot the computer, and create other not-so-pleasant effects. Viruses target areas such as executable files, program load files, device drivers (such as video) and in some cases, the Microsoft Normal.dot template.

How do these viruses enter a system? Diskettes traditionally represented the most common vehicle for intrusion. Hackers now introduce viruses via e-mail attachments, software purchased from commercial companies, malicious employees, Internet sites, and other electronic sources. Most of these viruses are aimed at infecting workstations.

Keeping Them Out

Preventing viruses from entering your company's systems can be addressed systematically. The initial step requires policies that dictate security measures. Reduce the number of viruses that intrude. Minimize the damage caused by those that you discover. Incorporate the guidelines that follow as corporate policies.

Backup data regularly so that when a virus does hit the company, all of your data is not lost. In environments where users store most of their information on local computers, ensure appropriate backups. Some companies require that employees place all of their files on a network shared on a central server to help ensure daily backups.

Scan all backups prior to restoring information so that viruses are not introduced into the system.

Do not allow software (other than corporate software) to be installed on computers within your company. Often employees carry programs from home and install them on their workstation. Ensure that policies prohibit this activity.

Teach user awareness of potential virus risks and how to minimize possible infections. If people are aware of methods to keep viruses out of corporate boundaries and how to minimize the damages if a virus does appear, they will understand their responsibility in your environment. Users should recognize the signs of viruses. Institute a procedure for reporting potential threats or suspicions of viruses.

Use virus protection software to detect known viruses. It is also important to download or install updates continually. Several vendors offer virus scanning protection and detection software. Some of these vendors include Data Fellows F-Secure Anti-Virus (www.datafellows.com), Network Associates/McAfee (www.avertlabs.com), and Symantec/Norton AntiVirus (www.symantec.com).

Auditing

How do you know if you have been attacked? Tools incorporated into any operating system, firewall programs, and other third-party vendor services allow you to monitor information within a corporate infrastructure. In terms of security, determine who is accessing the network, how files are being changed on the system, and where there are excessive "failed login" attempts. In some cases, tools are just a starting point. You may also need to monitor physical forms such as access to the computer system, electronic data storage media access, and access logs. When conducting audits, weigh your resources (time, manpower, and money) against the damaging effects of an intrusion. Strive for consistency, frequency, and depth.

Some of the areas to audit include:

• Changes to directories and files

- System logs for information such as login failures, deletion of system files, and system security changes
- Activities involving sensitive data
- Changes in user account access control privileges
- Possibilities of physical intrusions, including computer access or backup media access

Monitoring tools allow you to determine whether or not an activity is successful. Examples include successful and unsuccessful logins, file and directory access, use of user rights, security policy changes, and system restart and shutdown. If there are consecutive unsuccessful attempts by users, there may be a problem. In addition, several operating systems and applications deny access to a user for a specified amount of time if login attempts have failed beyond the prescribed limit. These systems can also be set to require the administrator to reset the password when failures occur.

CERT

The Computer Emergency Response Team (CERT) represents one of the main agencies for Internet security. This organization was formed by the Defense Advanced Research Projects Agency (DARPA) in November 1988. CERT's mission is to work with the Internet community to facilitate its response to computer security events involving Internet hosts, to take proactive steps to raise the community's awareness of computer security issues, and to conduct research targeted at improving the security of systems.

CERT products and services include 24-hour technical assistance in responding to computer security breaches, product vulnerability assistance, technical documents, and seminars. In addition, the team maintains a mailing list for CERT advisories, and provides a Web site (www.cert.org) and an anonymous FTP server (ftp://info.cert.org/pub) where security-related documents, CERT advisories, and a variety of tools are available.

Visit the CERT Web site to view the latest security alerts, including viruses and network vulnerabilities. In addition CERT acts as a central location to report incidents relating to any corporate security problems.

In your quest to secure the computer systems of your company, a comprehensive plan with consistent implementation is your first, and best, line of defense. Your goal is to prevent a hacker from entering or affecting your network, without limiting the communication between your employees and your customers. If you lack inspiration, take another look at www.alt2600.com, a description of hacker attacks and their weapons of choice. If you lack direction, review the policy and procedure instructions throughout this chapter.

Use any of the security measures available to you: user procedures, access restriction, passwords, firewalls, proxy servers, encryption, digital signatures, and audits. Determine which measures serve your goals by weighing their benefits against the cost of implementation and maintenance. Keep in mind that security is only as good as your governing policies and procedures, enforced so that everyone is aware of risks and remains accountable. Hardware and software deployments alone are insufficient. Make the commitment to secure your company.

10

Auction Technology

- Benefits of auctioning goods and services
- Types of auctions
- Exploring auction markets
- Establishing auction settings
- Reviewing vendor solutions

Auctions are common in the business world for negotiating trades of large monetary value. But consumer sales and small scale purchases have typically relied on the fixed price model, perhaps due to the high overhead associated with the auction or brokerage method. With Internet-based technologies, the high cost is nearly eliminated. Any product or service can be presented online to be auctioned. Items such as collectibles, toys, clothing, airline tickets, sporting event tickets, automobiles, and even houses are auctioned today via the Internet. How can you use auction technology in your business to increase revenue, sell surplus goods, replace traditional sales processes, and attract customers to your site?

The auctioning of goods and services incorporates several basic methods of determining prices and winners. Should you use open cry, sealed bid, or Dutch rules? The most common method, open cry, describes the auction of goods with the highest bidder winning. Is open cry the best method of auctioning your goods? Understanding auction technology and how to deploy it as a business process within your company can help you select the appropriate method and implement it efficiently.

How do you typically set prices for your goods and services? Businesses, both large and small, can now sell goods and services at market value. The concept of market value results from the law of supply and demand. You can manipulate the law of supply and demand to your advantage with auction technology. Would you let your customers set your prices? Before you respond negatively, remember that you can establish the timeframe and the minimum price acceptable to your company.

This chapter provides an overview of auction technology by considering its benefits, by describing auction classifications, activities, and markets, and by explaining software settings. It eases your selection process by surveying major auction vendors and by including sources for further research. The chapter also explains how to sell your goods online either at existing online auction sites or by integrating auction technology packages into your e-commerce environment. With estimates of auction sales comprising 10% of all e-commerce transactions by the year 2002, it is important that you consider auctioning as an alternative (or complementary) method of selling your goods and services online. Do not allow your competition to outbid you.

Benefits of Auctions

What does it cost you to distribute your goods and services? Could auctioning them off reduce that expense? Does your company generate surplus inventory that costs you money to maintain? Auctioning represents a method to reduce those inventory maintenance costs while distributing your inventory at a profit. How do you determine prices that ensure your profit while turning over products and services quickly? Again, auctioning may be the answer. Auctioning allows your customers to determine the price of the item or service. Use this arrangement to your advantage. Instead of limiting your sales to the consumer market, the auction model generates enhanced transactions between businesses.

When properly implemented, the benefits of auctioning include reduced distribution costs, reduced surplus inventory, and market pricing based on supply and demand.

Reduced Distribution Costs

Consider the time and expense involved in preparing for a traditional sale. Typically, when a company conducts a sale, goods are sent to a central location for inventory and tagging. And then the goods are redistributed to stores throughout a region (or nationally). Even if items are not centralized initially, sales personnel may still spend hours placing sale tags on the inventory. Certainly the world has left behind this inefficient practice.

Why not locate your entire inventory in a single location and auction it off online? This system of pricing and distribution offers several advantages. First, it reduces the time spent packaging the goods to be shipped to the central location for markdowns and then redistribution to the stores. Second, it reduces shipping costs. Third, it eliminates wear and tear (more specifically, loss or breakage) associated with the shipping process. And finally, this sales method encourages your customers to determine the value of offerings.

Reduced Surplus Inventory

Consider the beauty of selling every item or service you are able to offer. Now, think about how many businesses have surplus inventory or damaged inventory sitting in warehouses or on docks or locked away in storage. Which model represents your company? Electronic auctions may someday replace the way businesses currently mark down the price of merchandise, clear inventory of seasonal items, liquidate inventory of discontinued product lines, or sell damaged goods. Often, auction profits far outweigh the long-term expense of waiting for a better way to eliminate inventory.

Established Market Value

How do you determine the price of your product or service? How often are you wrong or wrong enough to create excess inventory? During typical sales, items are marked down by 20%, 30%, eventually 80%. Electronic auctions replace current practices of matching price to demand behavior (and their resulting markdowns) to ensure sales volume. In addition, auctions speed the turnover of product. Customers determine the cost of an item and the quantity to buy. Your task involves implementing and managing the auction process to your advantage. When you combine the benefits of reduced distribution costs, reduced inventory, and market valuations for your products and services, the auction model proves very compelling.

Auction Classifications

Auctions generally fall into the categories of open cry, sealed bid, Dutch auction, and second price, based on the rules governing them. Depending on the specific product or service you want to auction, select one of these four methods. The following sections define each auction type, explain how you can incorporate it into an e-commerce environment, and provide examples of successful implementations.

Open Cry Auction

Open cry auctions are what most people imagine when they hear the word "auction." Starting at zero or a specified minimum bid, bidders continue to offer more money for an item until the final minute on the last day of the bidding period. Then the customer with the highest bid wins, and purchases the item. This type of auction works well when companies auction goods whose prospective buyers can afford the time to place counter bids throughout the specified period and feel comfortable about making counter offers within a few minutes or hours.

Open cry auctions have worked extremely well for selling antiques, collectibles, books, tickets, business and office supplies, clothing and accessories, computers, electronics, cameras, toys and games, and other items that people feel comfortable valuing (by making repeated bids) in a short amount of time. Major online auctions sites, such as Yahoo! and eBay have been extremely successful in this arena, both from the perspective of the customer and from that of the site owners.

Sealed Bid Auction

A sealed bid implies that a bidder for the product or service is not aware of the other bidders' information, such as their names and how much they are offering. When would you employ this type of auction? Sealed bid auctions are practiced when it is not possible for the bidders to prepare counter bids efficiently. This could be because it takes time to prepare a counter bid. Sometimes the prior bid information needed to prepare the counter bid can not be disseminated to the other bidders instantaneously, or the bidders are not available to participate in the auction at the same time or the seller does not want other bidders to see counter bids. For example, imagine several vendors are bidding for a hardware contract within a company. Typically, companies do not like to reveal the competitors' information. They do not want vendors to base their pricing on other bids because it artificially influences the price. Sealed bids may work in this situation. Another example is a request for proposals in government-based auctions. It is to the seller's advantage to keep the bids private until the auction is completed.

Dutch Auction

Dutch auctions work by initially placing a good or service at the highest bid. The auctioneer then decreases the bid amount until the first customer bids for the good. The Dutch auction has a start date but does not have an end date. The seller has already placed the good or service at the highest price and attempts to control the final price of the goods.

This type of auction works well when buyers feel comfortable about making a bid in a few seconds and when you (as the seller) have a few expensive items with a higher demand. The short time frame is important because the first person that makes the bid wins.

Second Price Auction

In a second price auction, the bid amount represents the maximum amount the customer is willing to pay for an item, and not necessarily the current bid. The auction software determines the customer's current bid based on other bids in the auction and the bid increment. The bid will never be greater than the customer's original bid amount, but it may be less. Second price auctions work well when the item for sale is offered over an extended period and bidding is expected to be slow. From the buyer's perspective, second price auctions allow customers to place a maximum value on an item, bid that amount, and then pay little attention to the auction.

Preparing to Auction

As a vendor of goods or services, preparing to auction requires establishing the rules and conditions of the process. Although the preparation process is not complicated, the success of the venture results directly from the decisions you make prior to the process. The first decision concerns buyer/seller registration. In most cases, particularly if you are using the services of a host company, the registration rules and procedures are already established. If you are hosting your own auction, the auction component of your electronic commerce software provides registration templates for each involved party. Auction items and their descriptions are then incorporated into the process. In order to establish the temporal boundaries of the auction itself, set a start date and an end date (with time of day, if appropriate). Determine which auction method, open cry (with an opening amount), sealed bid, or Dutch auction (with highest starting price), most effectively addresses your needs.

Another important consideration involves delivering your products to the customer. Consider the demands of your potential customers when selecting a delivery vendor. These demands typically pit price against length of delivery period. United Parcel Service, the United States Postal Service, and Federal Express dominate the delivery scene, but other vendors are available.

Choose payment methods. Determine the auction closing rules and the notification process. Trade settlement configuration handles the payment to the seller, the transfer of goods to the buyer, and if the seller is not the auctioneer, the payment of fees to the auctioneer. Precise details of this process are provided by the hosting company or within the auction component of your e-commerce software.

Auction Markets

Auction technology has attracted a lot of public attention recently, driven by the exploits of eBay, Amazon.com, Yahoo!, and Priceline.com. These companies use a variety of auction models to sell goods and services to consumers via Web sites. A majority of the current auction activity, at least online, is between individuals. These auctions resemble garage sales when you consider the variety of merchandise and the diversity of participants. As dominant as the consumer-to-consumer activity is currently, the big jump in auction activity will be driven by the business-to-consumer and business-to-business markets. As you consider which auction method to employ (open cry, sealed bid, Dutch, second price, or some variation), remember to take advantage of the particular model, whether consumer-to-consumer, business-to-consumer, or business-to-business. Each of these market models, and the conditions that shape them, are discussed in the following sections.

Consumer-to-Consumer Auctions

Consumer-to-consumer describes a category of auction that encompasses the transaction (purchase/sale) of goods and services between individuals. Typically the entire transaction takes place on a hosted Web site, such as those of eBay, Yahoo!, and Amazon.com. An individual "places" an item to auction on the hosted site. When placing an item on the site, the customer must first select the appropriate category for the item. For example, users of Yahoo!'s auction services select among the categories of antiques and collectibles, arts and entertainment, books, musical instruments, tickets, business and office, clothing and accessories, jewelry, computers, electronics and cameras, home and garden, sports and recreation, toys and games, trading cards, and other goods and services. Potential buyers select a category, specify an item, and then bid for the item. At the conclusion of a specified auction period, the high bidder purchases the item.

Most consumer-to-consumer auction sites do not dictate how the final payments should be made. These online auction houses allow the buyer and seller to negotiate payment methods and shipping details. Typical arrangements include personal or cashier checks, credit cards, or money orders. While a majority of transactions conclude satisfactorily for both parties, fraud is still a possibility. One company that helps combat this problem is i-Escrow, a San Mateo startup, that holds the money and transfers it to the seller only when the buyer gets the merchandise. i-Escrow gets paid by taking a cut of the purchase price of 5% or less, depending on the total price. The company facilitates deals ranging from $10 to $10,000, with an average of $300. The president of i-Escrow estimates that roughly 10% to 30% of consumer-to-consumer auction transactions typically fail. However, faulty transactions handled by his company account for only 2% of the total. A neutral third party, such as i-Escrow, may further legitimize auctions as a compelling transaction medium.

Why does the consumer-to-consumer auction process work so well? The answers are rather predictable. Participants get excited during the bidding process, develop a sense of urgency as auctions close, and always love a potential bargain. If you have ever attended a live auction or bid for an item online, you have experienced the excitement that the event produces. Whether the item is nearly priceless artwork or third-rate samples, the tension and suspense holds its audience. Even when a bidding war concludes, another bout waits in the wings.

The limited quantity of a specific product, whether ten or a hundred pieces are available, and the limited duration of the bidding period, combine to create a sense of urgency among audience members and potential bidders. Remember the times you have observed (or participated in) infomercials on QVC. The piece count representing the number sold and the timer (counting down a particular offer) generate interest in the process. No one wants to miss out on a bargain. As a merchant, use this emotion to your advantage. Think carefully about the number of items to offer and the duration of the auction period.

The third reason that consumer-to-consumer auctions compel their participants is that everyone loves a bargain. With initial prices (bids) starting at zero or a specified minimum, the tendency is to believe that the resulting price does not reflect the entire value of the item. In addition, the winner of an auction relishes the feeling of having defeated his opponents, even if the price became unexpectedly high. The atmosphere and sense of power draw participants to online auctions. To get a taste of the auction process visit eBay. Browse the available categories and products. Spend some time watching an auction.

While Yahoo! offers auction listings at no charge to the seller, eBay does charge a fee to list items at their Web site. The two types of incurred fees are insertion fees and final value fees. eBay lists all of the rules and costs for listing products and services on their site. In order to list an item, visit the eBay site at www.ebay.com, and select the Sellers category. When you list your item for sale with eBay, you are charged a non-refundable insertion fee. The fee runs from $0.25 for inexpensive items to $2 for items selling for over $50. Insertion fees for vehicles cost $25, while those for real estate cost $50. At the end of an auction, you are charged a final value fee, based on the final value of your item. Final value fees total 5% of the amount of the high bid up to $25, plus 2.5% of that part of the high bid from $25.01 up to $1,000 plus 1.25% of the remaining amount of current high bid that is greater than $1000. Final value fees for vehicles are fixed at $25. Real estate does not incur a final value fee. Calculate your total commission price to determine whether these fees seem excessive, and then shop the competition.

We have established the participant's level of interest and explained what it costs to conduct an auction. How do you set up a consumer-to-consumer auction? Take a look at the process for the Yahoo! site (www.yahoo.com), where items can be placed online for free. Just follow the steps below:

1. Select a category.
2. Upload a photo of your item.
3. Enter the title and description of your item.
4. Enter shipping information.
5. Choose payment methods.
6. Establish the quantity available.
7. Set your starting and reserve price (optional).
8. Determine the duration and closing time of the auction.
9. Review your selections.
10. Submit your auction settings.

Most online auction sites incorporate a process similar to this one for setting up a specific consumer-to-consumer auction. Details about the process and support information are listed at each site. A Forrester Research report, dated March 1999, reported that the consumer online auction market would grow to $19 billion by 2003, fueled by the growth of consumer-to-consumer and business-to-consumer players. Consumer-to-consumer auction sites gain popularity as their benefits are understood and appreciated. eBay and Yahoo! are the current market leaders and offer the most exposure, a critical factor in effectively selling.

Business-to-Consumer Auctions

Jupiter Communications stated that online retailers will continue to have difficulty anticipating consumer demand for purchasing specific items online. As they look to liquidate excess inventory, retailers should embrace the inherent interactivity of the Web and move beyond traditional fixed-price discounting to online auctions. In addition, online retailers that take advantage of auctions benefit from higher repeat visits, more frequent purchases, and greater promotional value.

Jupiter Communciations' research also shows that 1999 is the year that business-to-consumer auctions move beyond limited technology product offerings to a more diverse product mix. More technologically-savvy users and the availability of other merchandise, including toys and apparel, should attract a more mass-market consumer to the interactive sales format.

Most auction sites rely heavily on using traditional advertising venues until their sites are known for providing auctions. A major department store wanted to know how to inform customers that they are auctioning items online. The answer was to use traditional marketing. They placed advertisements announcing a sale, but also listed the Web site and explained the auction format. Instead of requiring customers to drive to the store, the department store allowed customers to visit online to bid on items. By conducting weekly auctions, the store encouraged visitors to become frequent shoppers and to buy products that might otherwise be overlooked.

Business-to-consumer and consumer-to-consumer sites perform a valuable service for direct marketers. For example, imagine that you create necklaces. Most of your sales are generated through classifieds advertisements or trade magazines. You could set up an auction online to sell your product. Using an auction to generate sales does not prohibit you from using your traditional channels. The risk of the new arrangement is very low. When you set up the auction, just set a reserve price high enough to cover your costs. A reserve price is the lowest price you are willing to accept for your product. Perhaps auctioning will eventually influence your marketing expenses and your revenue positively.

Business-to-Business Auctions

Think about how many companies have excess equipment, parts, or any other products sitting in storage, requiring a sizable maintenance expense. Business-to-business auctions are used effectively for reducing this inventory, liquidating discontinued product lines, selling damaged items, and selling surplus goods. In addition, as auction technologies continue to improve, look for more companies to sell products and services to other companies. In some cases, the revenue gained by auctioning off the inventory may even outweigh the cost of maintaining the inventory each year.

Auction Settings

Auction settings control the behavior of an auction and the amount of information about the auction that should be disclosed to a customer. Avoid changing auction settings while an auction is in progress because the results are often unpredictable. For example, if an auction is changed from sealed bid to open bid, customers do not see the bids displayed on their browser until after they refresh their screens. Auction settings, generally available across the range of software solutions, include bid increments, minimum and current minimum bids, sealed bid and inventory settings, reserve price, dates, and tie resolution.

Bid Increment

The bid increment is the minimum amount a bid must be raised to place a bid on an item. Set a bid increment to encourage customers to increase their bids by reasonable increments. In second price auctions, the bid increment is used to determine the amount a winning bidder will pay for an item. The bid increment is used with a reserve price to encourage customers to bid on an item at least until meeting that price.

Minimum and Current Minimum Bids

The minimum bid is the lowest starting bid that you agree to accept for an item. Once an auction has begun, the current minimum bid determines the lowest bid a customer can place on an item. For winning price and clear price auctions, the current minimum bid is the amount that must be bid in order to win an item. When only one item is for sale, the current minimum is the current high bid plus the bid increment. When more than one item is for sale, the current minimum bid is the amount that must be bid to win at least one of the items. For a second price auction, the current minimum bid is based on the winning customer's current bid. Otherwise, the current minimum bid is calculated the same as winning price and clear price auctions. In reserve price auctions with multiple items, the minimum bid increases by the bid increment until at least one item has been purchased. After this reserve price has been met, the current minimum bid will be either the reserve price (if more than one item is available), or calculated the same as the winning price and clear price auctions. For sealed bid auctions, the current minimum bid is typically not shown. Take caution when displaying the current minimum bid for sealed inventory auctions because customers could use this information to guess the number of items available.

Sealed Bids

The sealed bid setting should be selected to limit information about the number of bids on an item, including the amount of bids, as well as the identity of bidders. While the sealed bid auction is running, very little information should be disclosed to customers. No bid amounts, nor any information that allows customers to guess bid amounts, should be disclosed. This includes information such as the current minimum bid, the bid price, and the number of items won. In addition, sealed bid auctions do not typically reveal any information about the bidders, such as bid names, e-mail addresses, and the number currently won. Typically, a sealed bid auction displays the minimum bid (not the current minimum bid), and the number of items available, assuming the auction does not have a sealed inventory. When the auction is complete, the site designer may disclose the results of the auction. Whether or not to disclose auction results depends on the site policy.

Sealed Inventory

The sealed inventory setting should be selected when you want to withhold the number of items available. For example, if the seller has 10,000 items available, but only expects a few hundred bidders, choosing this setting encourages customers to bid against one other, which can result in a higher price for each item.

In a sealed inventory auction, the number of items for sale is not disclosed, nor is the number of items won, since this may allow customers to guess how many items are available. A sealed inventory auction may disclose bid amounts and customer names depending on site policies and the sealed bid setting; however, the number of items each customer won would not be disclosed.

Reserve Price

A reserve price is the minimum amount for which the seller is willing to sell an item. Typically, customers should be informed that the auction is a reserve price auction, as well as when the reserve price has been met. However, displaying the actual reserve price once it has been met is optional. When bid amounts and numbers won are displayed, the reserve price is often inferred.

Dates

The start date of an auction is when customers are first able to place bids on an item. After the end date, bids are no longer accepted and the item becomes available for the winner(s) to purchase. When creating an auction, set the start date less than the end date or an error occurs. The start and end dates of an auction typically display for the customer.

Tie Resolution

An auction may result in a tie. In order to resolve the tie, set tie resolution rules. The tie-breaking criteria could include bid price (almost always first), the number of items requested, and the date/time the bid was placed. The first tie-breaker that applies determines a winner. For example, in an winning price auction with five items available, customer A bids $10 for three items. Customer A receives a message saying three items are being won. If customer B then bids $10 on four items, customer A will receive a message saying one item is being won.

Vendor Solutions

To implement auction technology for any of these three market models, you can select solutions from four dominant vendors: Microsoft, Netmerchants, OpenSite Technologies, and IBM.

Microsoft

Microsoft (www.microsoft.com) offers an auction component that supports three types of auctions: winning bid, clear price, and second price. The auction type, once selected, controls the method used by the auction component to determine the winning customer, and the amount the winning customer must pay.

A winning bid auction allows winning customers to receive the item they desire for the amount bid. This type of auction works well if few items are on sale, many customers bid on a single item, and the auction is open for only a short time. If there is more than one item, then winning customers receive the number requested at the amount they bid. For example, suppose there are two items at auction. Customer X bids $10.00, customer Y bids $5.00, and customer Z bids $3.00, and each customer requests one item. Customer X receives one item for $10.00 and customer Y receives one item for $5.00. However, if customer X requested two items at $10.00 each, both items would be awarded to customer X.

Winning customers in a clear price auction receive the item they desire for the lowest winning bid. This type of auction works well when many items are on sale and you want all winning bidders to pay the same price. For example, suppose there are three customers bidding for only two items. Customer X bids $10.00, customer Y bids $5.00, and customer Z bids $3.00, and each customer requests one item. Customer X and customer Y each receive one item for $5.00.

In a second price auction, the auction component places bids for the customer. Each time the auction component places a bid for a customer, the component increases the bid by an increment, which is defined by the auction manager. In a second price auction, each winning customer pays the lowest of the following: the highest losing bid plus the bid increment, the lowest winning bid, or the reserve price, assuming the number of bids above the reserve price is less than the number of available items. For example, suppose three customers bid for two items and the bid increment is $0.50. Customer X bids $10.00, customer Y bids $5.00, and customer Z bids $3.00, and each customer requests one item. Customer X and customer Y each receive one item for $3.50. Each of these three types of auctions address a different buyer audience and should be employed accordingly.

Netmerchants

Netmerchants (www.auctioneer32.com) offers Auctioneer, an easy-to-administer and easy-to-install auction software package, especially for those sites that focus on the consumer side of auctioning. Auctioneer's modular design and flexibility allow you to clearly define your intended auction audience. For example, you can set up your own auctions to sell your merchandise, or you can create an auction community for users to sell to other users. Both are equally supported, but you should make the decision early in the configuration process since the package defines your setup, such as billing and user options.

There is a high level of control over the actual bidding process. You can define bid increments based on different criteria. For example, you can have higher increments for more expensive items and lower increments for less expensive items. If you are offering open auctions, users can leave detailed feedback on other users, letting the public know about the reliability and integrity of sellers and buyers.

Auctioneer's variety of features should appeal to most users. A chat room allows users to post messages to the user community, while classifieds allow for items to be sold without the auction process. With Compubid, an automatic bidding feature, the computer bids on behalf of the seller until the reserve price is met. A search engine allows users to find items quickly, while sellers can see an overview of their performance and the types of auctions that work best. For higher-end sites, Netmerchants is developing Auctioneer SQL, which is integrated with heavy-duty SQL databases (as opposed to the more modest databases that are accessed via ODBC).

Auctioneer incorporates particularly rich e-mail functionality. Users can sign up for e-mail catalogs of items within chosen categories, allowing them to browse the products and offerings at their leisure. In addition, a personal shopping system notifies users when new items have been added to specified categories. Auctioneer offers templates and configuration files for managing all aspects of e-mail, including end-of-auction notices, registration confirmations, bid confirmations, and outbid notices. E-mail addresses that are posted on the site can be protected to prevent harvesting from spammers or competitors.

OpenSite Technologies

OpenSite Technologies (www.opensite.com) offers auction technology in three flavors, depending on how comprehensive a solution you require. The OpenSite editions include OpenSite Auction Professional, OpenSite Auction Merchant, and OpenSite Corporate edition.

OpenSite Auction Professional edition is an entry-level version designed for beginners on a limited budget. Only the owners of a Web site can offer items for auctions in this version. It includes a complete set of Web pages (including home page, registration page, contact page, help page, category and item pages, auction results and more) that can be directly incorporated into an existing Web site or serve as a basis of a new Web site. Included is a product search engine, credit card verification (without transaction capabilities, just validation), and an invoicing mechanism to automatically calculate invoices, including charges for taxes, shipping, fees, and commissions.

OpenSite Auction Merchant edition is designed for merchants that want to build auction capabilities into existing Web sites and offer auction capabilities to their users for a fee. In addition to the capabilities found in the Professional edition, the Merchant edition includes support for banner advertising, consignment auctions, and a fixed-price electronic commerce storefront.

OpenSite Corporate edition is a turnkey solution for auction-centric sites, allowing program development via Microsoft's ISAPI and integration with an Oracle database. Also included are private auctions and online credit card payments. The latest version of OpenSite adds a lot of flexibility in auction formats, especially by supporting reverse, modified English, sealed bid, private, and consignment auction variations. Reverse auctions allow buyers to request goods or services and then companies bid for those sales until a final bid has been reached. Industry experts believe that reverse auctions will play a significant role in business-to-business transactions. When multiple quantities of the same item are offered for auction, conduct a modified English auction. All winning bidders pay the amount of the lowest winning bid so that no one pays more than anyone else for the same item. If you need to permit only specific bidders to participate in an auction, conduct a private auction. And if you want to run an auction for other sellers (in order to collect a commission for your work), conduct a consignment auction.

IBM

IBM (www.ibm.com) offers a new auction package that can support many different types of auctions including open cry, sealed bid, and Dutch auctions. Each type of auction can have many variations such as reserve prices, information available to the bidders, and tie breaking rules. The user is solely responsible for the type of auction chosen, the rules and their administration, as well as complying with all applicable legal requirements. The package provides those functions to auction the products listed in the Net.Commerce Catalog. For more information about their Net.Commerce product, see the description in the Vendor Solutions chapter. The auction package is currently available as an early release technology preview for limited internal evaluation and testing with IBM Net.Commerce (version 3.1 and 3.1.1 Windows NT), in accordance with the associated license agreement.

A forms-based interface is available to merchants to select a product from the catalog and place it at auction. The forms allow the merchant to specify the type of auction being conducted (open cry, sealed bid, Dutch), the parameters negotiated (price, delivery dates, terms of payment, etc.), the starting date and the time of the auction, and the auction closing rules. When setting up the auction, the merchant can also specify rules to control the pace of bidding. These rules specify the starting bid and bid increments.

Buyers have a graphical interface to search for auctions and place a bid in an auction. A deposit can be taken when the bid is submitted. The deposit is taken online using the payment mechanism available in Net.Commerce to buy fixed price items. The buyers can view their past bids and increase them. In open cry auctions the buyers get regular updates of current highest bid. Auctions can be closed automatically at a specified time. Open cry auctions can also close automatically after a certain period of bidding inactivity. Auctions can be closed manually by the merchant at any time. At the close of auction, shipment of the merchandise and collection of payment balance is accomplished using existing Net.Commerce order mechanisms. If you use

Net.Commerce as your e-commerce solution, consider the benefits of incorporating IBM's auction technology.

Auction technology allows your business to increase revenue, sell surplus goods, improve traditional sales channels, and attract more customers to your site. Your understanding of the various auction classifications and the benefits of each makes it easier to select a vendor's auction package. Once you have made this decision, review the descriptions of auction software settings. Then turn to the next chapter to begin the deployment of your e-commerce solution.

11

Project Deployment

- Creating a project deployment plan
- Establishing the right vision and goals
- Developing your e-commerce strategy
- Deploying the solution
- Staying focused

Now it is time to take the big step. Deploy your electronic commerce solution. You have waded through our book, various magazines, online articles, conversations with experts, and lots of other sources to gather ideas, strategies, instruction, and inspiration. How do you finally get started?

The best way to develop an e-commerce strategy is to approach the challenge from a business perspective, prior to even asking the technology-based questions. It is critical that the designed solution meets business goals and processes, and most importantly, provides a substantial return on your investment. The focus is on design, development, and then deployment. Define the elements that constitute each step, and remain focused on their order.

This chapter functions as a high level overview of your approach to an electronic commerce project. The eight steps outlined below lead you through the creation of your business goals and strategies, from your current process to a revised process, to revising your architecture, and finally to designing, testing, and deploying your e-commerce solution. Follow these steps as a recipe for your project.

1. Define your vision and goals.
2. Develop your e-commerce strategy.
3. Move from current process to new process.
4. Move from current systems architecture to new systems architecture.
5. Design the electronic commerce application.
6. Pilot and test the system.
7. Deploy the system.
8. Focus on success strategies.

Whether your business sells products to other businesses or to consumers, you need to approach e-commerce as more than just computerization or a way to electronically enable the enterprise. E-commerce must be approached with the core components of selling products and services, providing customer support and services, marketing, sales, and communication with distributors, suppliers and other trading partners. These are the standards of any business, whether online or traditional.

Developing an e-commerce solution does not mean throwing away all of your current methods of doing business. It means expanding the way you think about your business and expanding the way you generate your profits. It means gaining additional customers through globalization. It means removing barriers of entry and not being confined by limited resources. It means reducing costs associated with ordering goods from your suppliers through supply chain management initiatives. The results leverage your current efforts and existing architecture to expand your business with an online solution.

According to Gartner Group, "Enterprises that have implemented electronic commerce as part of a business strategy have been more successful in reducing business cycle times, improving cash flows, reducing inventories, decreasing administrative costs, and opening new markets and distribution channels." Sound promising? Take the next step by deploying your e-commerce solution.

Defining Your Vision and Goals

Defining the vision statement and goals for the your e-commerce solution keeps you focused and pointed forward. For example, when we started our company, we stated that "We are the e-commerce source worldwide." Our goal was to become the source of knowledge for the world on how individuals, entrepreneurs, small, medium, and large companies could successfully deploy an e-commerce solution. We represent the eyes and the ears of industry in determining how companies have and have not been successful in e-commerce. We harness business and technology trends and development methods to implement e-commerce solutions based on our industry experience. Our strategy focuses on delivering our knowledge to both businesses and the technically-savvy through university courses, technical courses, company seminars, an e-commerce-related book series, and involvement in industry.

Vision Statement

Your vision statement sets the stage for the entire e-commerce project. Vision statements ensure that the technical solution meets the overall business vision. Think big! Think big picture! Once you have established your vision statement, the next step is to determine how to get your company aligned with this vision. The most efficient alignment results from a solid definition of your goals and strategies.

Goals

How will you become the e-commerce leader? What are you trying to accomplish? Where do you want to be in a year and in five years? What are your goals? The goals you set should be measurable. Did you do better than last year? How do you know? Often companies set goals that they cannot measure. Without some measurements, success proves nebulous. Create a series of measurable goals. For example, pretend that you have deployed an e-commerce site to sell jewelry. Your goals might include:

- Each quarter, increase revenue by 30%, constantly promoting the site, thinking globally, and providing an extremely user-friendly experience.
- Become the leading source for selling jewelry online by developing vendor relationships with the five top jewelry suppliers.
- Provide the highest quality of customer service with 90% or higher satisfaction rates by focusing on quick and friendly customer service, online support, and 24-hour phone support.
- Ensure repeat business of at least 20% by taking a proactive role in servicing the customer, providing the highest quality products, and target marking.
- Become the leading provider of jewelry in the United States, South America, and Canada by localizing sites to the top five counties and languages.
- Decrease distribution costs by focusing on efficient methods of packaging and delivering goods to the consumer.

These goals should be reviewed on a quarterly basis. The first few months of operations establish a pattern for your long-term success. If you exceed your expectations, adjust. A computer hardware supplier, Compaq, realized one million dollars a day in sales shortly after going online. They did not anticipate such positive results in such a short period. Learn a lesson from Compaq. Be prepared for success by selecting the appropriate strategies.

Developing Your Strategy

Developing an e-commerce strategy for your business requires the dedication of your staff. Allocate your resources (time, money, and materials) efficiently for this project. Remember that e-commerce is not only an IT function, it is a critical element of your core business. Define your key players and keep them involved.

This section reviews methods for determining and defining those e-commerce strategies that deliver the greatest value in the shortest amount of time. If you are new to e-commerce, do not try to implement the entire solution initially. Keep the project small and manageable. And then create infrastructure capable of supporting your growing business. The techniques for developing new strategies include researching the markets, brainstorming, prioritizing, and creating a scope statement.

Researching the Markets

Prior to determining where your company should focus its efforts, spend some time researching your industry and your competitors. Include in your research any team members that will contribute to the definition of your online niche. Determine the minimum requirements for your e-commerce solution. Your goal here is to create a solution that ultimately differentiates you from your competitors and improves upon their offerings.

Most research can be conducted via the Internet. Go to your competitors' sites and determine what they are providing their customers. Catalog not only the product or service offered, but the components of the experience: customer service, delivery options, alternatives, links to other sites, etc. Visit vendor sites, such as IBM, Microsoft, and BroadVision. Review the customer profiles sections of these sites to learn how they are successfully deploying e-commerce solutions. Another great research method involves visiting online magazines, such as Business Week or Forbes, and searching on your industry to see how it applies e-commerce solutions. The more information you gather about your competitors, the more effective your brainstorming will be on new and innovative ways of doing business online.

Brainstorming

Prior to defining the new process of doing business, spend some time brainstorming on all of the areas where you could use e-commerce within your company. This brainstorming should be performed as a group exercise with all of the key players in the company, including marketing, sales, product specialists, business personnel, and MIS professionals. These sessions should also be professionally facilitated and documented to ensure that all of the ideas are realized and captured. Even if the ideas seem way out of reach, you may be able to use them later. In addition, do not consider the technical feasibility at this point. Think purely about the business. Too often companies think about the technology and limit their ideas accordingly. Areas that you should focus on vary, depending on whether you are selling products and services or attempting to streamline the business process through supply chain management.

If you focus on selling products and services with a business-to-business or business-to-consumer e-commerce solution, the goal is to create a user-friendly site where catalogs and service areas are easy to navigate. Marketing, personalization, and customer service each play a major role in determining what business concerns to include.

If supply chain management between businesses is your focus, place emphasis on streamlining the business processes to reduce costs, become more efficient, and improve distribution channels. In supply chain management, the goal is to seamlessly interface with your vendors' or suppliers' existing business processes. However, you should also be thinking about how to provide additional services to the companies with which you do business. In some companies, especially larger ones, both areas should be addressed.

When facilitating or actively participating in a brainstorming session, consider your answer to these questions. They ensure that you keep your energy focused on business. They capture the main areas required to determine your strategies.

• What are the major projects of interest to your company?

- What are your new revenue streams?
- What sales and marketing techniques should you use ?
- How should you provide and enhance customer service?
- What products and services do you offer?
- How do you ensure repeat business?
- What are the geographic and demographic characteristics of your customer base?

Spend some time defining what e-commerce projects are required to implement a fully functional e-commerce solution. Typically, your business is comprised of several concerns that lend themselves well to effective e-commerce solutions. For example, you may choose projects focusing on:

- Business-to-consumer solutions: selling to consumers
- Business-to-business solutions: selling to companies
- Supply chain management solutions: automating processing between trading partners
- Intermediary solutions: providing portal services for other distributors or vendors

Once you have reviewed these business opportunities, rank each one, and then determine which category adds the most value to your corporation within the shortest period. Begin it. Based on your resources, you can start the next highest priority project as soon as feasible.

Since you are entering a new market for selling goods, new methods may generate additional revenue. Some of these methods include:

- Advertising for other companies at your site
- Providing subscriptions to online information
- Partnering with companies to generate transaction fees by capturing a percentage of each transaction
- Selling directly to consumers or businesses
- Delivering digital content
- Collecting commissions for matching buyers to sellers

To maximize sales and marketing efforts, employ traditional and electronic methods of advertising. Determine all of the ways that you can attract customers to your site. Even if some methods prove expensive, the resulting revenue may justify your costs. Consider using many of the major opportunities discussed in the marketing chapter. These categories include search engines, advertising banners, online classified advertisements, message boards, user registration and e-mail, links on other Web sites, newsgroups, discussion lists, traditional media, press releases, trademarks and branding.

In order to ensure that customers experience your personal touch, look at all of the ways that you can provide great customer service. This area is a critical success factor for any e-commerce site. Customer service reassures, even more so than for traditional selling, as your customer may already be wary of purchasing online. Superior customer service draws customers back to your site following initial transactions. Offer services, such as FAQ listings, e-mail response messages after purchases, and 24-hour phone support to service your customers.

Determine what products and services of your business would prove compatible with an e-commerce solution. Some of the products you currently sell may not be a good fit. If you are in a service business, think of service that can be marketed, supported, or sold online. Often, people forget about the service category and instead, classify e-commerce as purely for product sales. Do not make this mistake. A variety of companies offer technical support, consulting services, and advisory information very effectively online.

It is also important to think about ways of ensuring repeat business. Think of all of the ways that you could create a commerce site that would keep customers coming back. Some of these methods include providing great customer service, making the site a new experience, providing quality products and services, and taking a proactive role in the relationship through push technologies or by sending personalized e-mail messages.

Who are your customers? Think of all of your markets by country, language spoken, and demographics such as age, sex, income level, and interests. What languages will your site support? Will you create different product lines by country? You may also consider localizing your site to different regions or countries. Do not treat the differences among your audience members as a burden. Instead, harness these differences to widen your potential pool of customers.

Prioritizing

After conducting the brainstorming session, start prioritizing each category and determining which areas you want to include and to exclude. List all of the brainstorming ideas and then review each idea. Prioritize the list. Brainstorming allows you to think outside of the box. Prioritizing lets you sort among the ideas generated and rank those that support your company's goals and strategies.

Creating a Scope Statement

Once you have prioritized each element, create your scope statement. The purpose of the scope statement is to break the project down into manageable units of work. Breaking the project down into these manageable units ensures the delivery of a quality e-commerce solution. You should create a scope statement for each phase of the project. The scope statement places boundaries around what you hope to accomplish. It includes a description of what elements are included in a project, and more importantly, what elements are excluded.

Considering the Business Process

After defining the scope statement, it is time to look at ways of reengineering the current process or method of doing business in order to enable the new e-commerce model. Capturing the current business process consists of reviewing and documenting how you are currently selling products and services. It is important to review the process because once you understand the current methods, you can work on defining the new methods.

Capturing Your Current Process

When capturing the current process, make sure to include each major process and the responsible individual (documented by title, not by name). Capture information such as who handles each step in the sales process and what media is employed (computer, paper, e-mail, etc.). The goal is to ensure that you gather all of the following information: form used, required fields or data, internal process built into the software, authorization and security requirements, as well as credit card transactions and paperwork that is delivered. For example, sales processes typically include the following considerations:

- How are employees selling products and services?
- How are employees tracking orders?
- How are goods being distributed to the customer?
- How is the company receiving inventory?
- What accounting takes place?

The goal is to gather all of the information being sent through the organization and determine who is responsible for each process. To capture the process, several methods may be employed. One form we use to capture the components of the process is a Roles and Process chart. If you know your current method of doing business, it serves as a basis for reengineering the process, while leveraging much of your current process. The example in Table 11.1 illustrates the roles involved in the processing of customer orders from the time the customer calls until the distribution of the goods.

Consider how concisely the Roles and Process chart captures this process. The process clerk picks up the phone and enters the order into SAP. Once the order is entered, it is processed in SAP and an order form is printed. The process clerk then sends the printed order to the warehouse office clerk. The warehouse office clerk gives the printed order to the warehouse supervisor who then selects the goods and packs them for shipping. The warehouse supervisor completes the shipping form and places it on the package, sending the inventory to the shipping department. At the same time, the accounts payable clerk prints a copy of the invoice from SAP and mails it to the customer. The customer, after receiving the inventory, pays the invoice and mails it back to the accounts payable department at the vendor site.

Table 11.1 Roles and Process Chart

Roles & Process	Order Process	Inventory Process	Shipping Process	Account Process
Customer	1. Places order			8. Pays invoice
Process Clerk	2. Answers phone 3. Enters order into SAP form 4. Prints order & sends to warehouse clerk			
Ware-house Clerk		5. Gives order to supervisor		
Ware-house Supervisor		6. Selects & packs goods	7. Ships goods to customer	
AP Clerk				4. Mails SAP invoice to customer 9. Receives payment

New Process Design

Redefining your business process includes ensuring that requirements stated in the scope statement are incorporated. One of the key factors for success is to ensure that the new process also aligns with corporate goals and strategies. As you redefine the process, make sure that you can match it back to your corporate goals. We practice this technique when conducting workshops for customers. After defining the new process, we ask the customers if the process is aligned with the goal stated earlier in the project. The result is a constant focus (on deliverables) that creates real value, instead of just "nice to have" features. Such features may be timely, but result in a very low impact.

The easiest way to document the new process is to write the information in a paragraph format. Once you have formalized the information, map it out in a diagram, like the one introduced in the previous section. At this point, you will probably notice some discrepancies or steps missing in the process. So sit with your group again and walk through the process enough times for the results to make sense. The more time spent ensuring that the new process is what you want, the less time it will take the e-commerce site developers to implement the technical side of the solution. The most time-consuming and costly part of a project often is having to redesign a system again and again when process is introduced or changed.

At this point, start evaluating which e-commerce vendor solution best meets your business needs. Consider the built-in workflows, features, and functionality. With the information gathered in the scope statement and the information gathered from the new process design, you should be able to reasonably select among the vendors. However, do not make the final decision just yet. You also need to consider your current systems architecture. You need to catalog what systems you currently employ and how to integrate with them.

Revising Your Architecture

After redesigning your company's business process to enhance the proposed e-commerce solution, focus on your current architecture. Consider all of the existing systems that require integration and then map a revised architecture. Even if you intend to outsource a majority of the development work, knowledge of your systems and any proposed changes improves the dialogue between the principals and the developers.

Current Systems Architecture

During the review of the current business process, it is time to start mapping the current architecture and systems. The goal now is to document all of the existing systems that are are part of the current architecture. Once you know the current architecture, you can determine what additional systems are needed and what commerce software to select.

Some of the major current systems to document include network operating systems, Web server software, relational databases, transaction-based systems, ERP systems, EDI systems, and all hardware capacities and platforms. In addition, if you are integrating with other businesses, document their systems that you need to integrate.

The resulting documentation should include a high-level architecture diagram and a description of the data required from each system. Remember that your initial work requires that you capture your current architecture. Figure 11.1 illustrates a high level snapshot of an overall architecture.

Figure 11.1 Current Systems Architecture

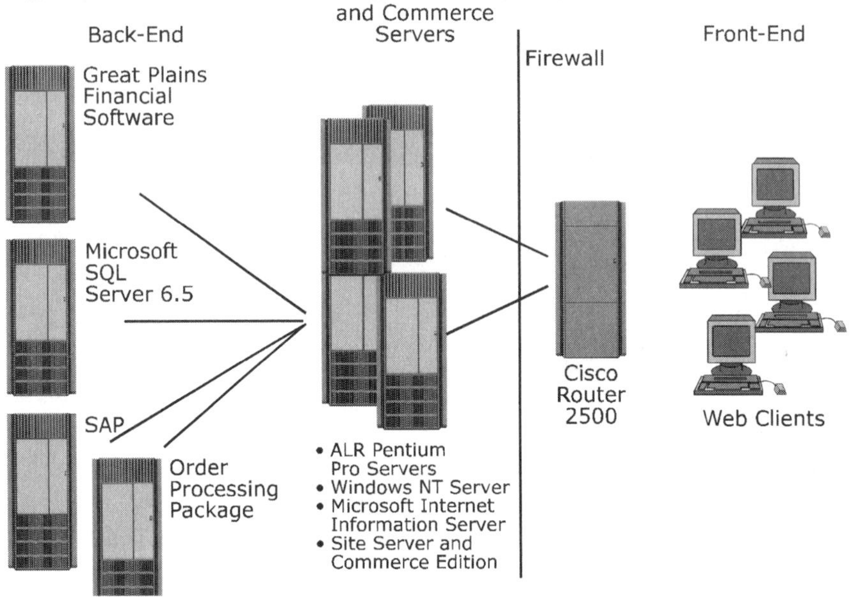

Back-End   and Commerce Servers   Front-End

Firewall

Great Plains
Financial
Software

Microsoft
SQL
Server 6.5

SAP

Order
Processing
Package

- ALR Pentium
  Pro Servers
- Windows NT Server
- Microsoft Internet
  Information Server
- Site Server and
  Commerce Edition

Cisco
Router
2500

Web Clients

New Systems Architecture

Based on the revised business process, focus on the new systems architecture. Begin by determining the tools necessary to deploy the solution. Specifically, you will need to determine what additional hardware and software is required. Back-end system requirements include additional servers and the capacity required, security systems, credit card verification systems, relational databases, imaging systems, and document management systems. Internet/commerce server requirements include additional servers and the capacity required, network operating systems, Internet server, commerce server, and security requirements. Front-end client details include workstation hardware and the capacity required, operating systems, Web clients, proprietary software, and security log-in requirements. Intranet/extranet architecture requirements include firewall security and proxy server information. Additional tool sets could include ISV plug-ins and back-end connectivity tools.

Figure 11.2 New Systems Architecture

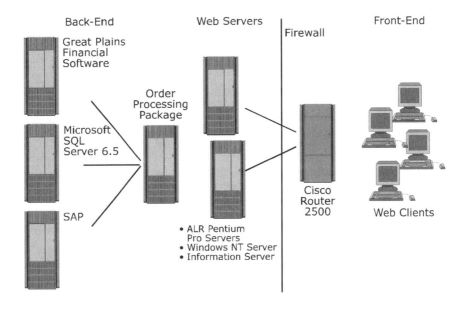

The difference now (illustrated in Figure 11.2) is that all back-end systems seamlessly integrate with the Web servers and the commerce server has been added. The business has leveraged existing systems to support a complete e-commerce solution.

Deploying a Solution

The remaining question is whether to build the solution yourself, host it yourself, partner with another company, join a mall, or outsource everything. The response to this question is primarily driven by cost, resources, technical expertise, and time. Each option has its advantages and disadvantages. The remainder of this chapter considers the range of deployment options and the characteristics of each. Specifically, it includes building and hosting within your company, developing internally and hosting at a commerce service provider, partnering with another company, joining a mall, and outsourcing the entire deployment.

Building and Hosting within Your Company

Building and hosting within your company offers several advantages, including a higher degree of control of the systems. In addition, you maintain more flexibility in making design changes, upgrading systems, tying into existing systems, and tying into your buyer and supplier systems for supply chain management. On the other hand, it is more difficult and expensive to deploy internally than to host with a commerce service provider. When building and hosting, your company must "rent" the right skill sets to build and maintain the solution. In addition, you need to purchase all of the hardware, software, and dedicated links to the Internet.

The following paragraphs walk you through the main areas you need to focus on when deploying an e-commerce solution internally. These areas include proof of concept, commerce application design, piloting and testing, and deployment.

Prior to developing the entire solution, you may want to ensure that all of the technologies selected integrate into existing systems, as well as into the new architecture. The purpose of the proof of concept is to integrate each component into the new environment and ensure that the site can connect to all systems successfully. A proof of concept should not take more than two or three weeks. This process allows you to avoid major code changes or drastic redevelopment if integration were to fail during deployment. Do not focus on creating a pretty site. Functionality is the key to this effort.

The application design phase consists of defining the details for the application. Roles and Process charts are used as a starting point for the application design, providing the workflow of the application. Once the developers study this information, they can start working on the detailed components of the e-commerce site design. Their components include functionality and workflow, security, detailed screen layouts, and graphics and interface requirements. The business and IT group should review the system design at least every two or three weeks to ensure the look and feel is user-friendly, easy to navigate, and provides the features and functionality specified in the requirements.

A pilot test, in addition to general application testing, is beneficial (and in most cases, required) after the application is developed and implemented on the existing systems. Test for usability and functionality. The usability test ensures that the potential customers can easily navigate through the site, that load time is minimal when opening pages through dial-up lines, and that the design reflects the enthusiasm you want to lend your customer. The functionality test is focused on system and application testing. All of the applications must be tested to ensure that components are working properly. Additional testing includes processing the number of transactions required when going online, integrating into existing systems, and testing performance.

Prior to testing, ensure that there is a documented test plan for the entire system. After each test, indicate those that were successful and those that did not pass (or were marginal). The tests that did not pass are then returned to the developers for further updates and improvements.

Having a pilot group provides additional feedback prior to going live on the Internet. The pilot group should be small and consist of at least one or two users from every group that will be using the new e-commerce site. Be aware of scope creep during this stage. Users of the system may start asking for additional features and functionality that may not be required and are out of scope. Document any additional requests. You may be able to include some of them in later phases of the project.

It is time to go live. In the deployment phase, the developed application is ported to the production environment. Ensure that all security is in place, such as Web servers, operating systems, firewalls, and proxy servers. Also ensure all of the standard settings are in place because the application has been moved to the new environment.

Hosting at a Commerce Service Provider

Another alternative to establishing a solution is to develop the site internally and host it at a commerce service provider. When using this type of host, you need to follow the same steps as building internally. The difference is that the testing and deployment is performed at the commerce service provider's site. This option provides control of software selection and design. And you are responsible for design updates. You do lose, however, some flexibility due to the distance between you and the host machine. Most merchants that outsource the hosting of their solutions mandate that the commerce service provider is in the same city as their business. Local commerce service providers tend to offer better customer service and more timely updates than national providers.

The best method for finding a provider is to first select the commerce vendor software solution. Then go to that vendor's site and find out what providers host that type of solution. In the Vendor Solutions chapter, we discuss sites that provide hosting services. For example, IBM solutions can be hosted with IBM, while BroadVision included a list of resellers that developed the commerce solution and hosted the solution. You also should contact current customers of these services to ensure that they provide adequate customer service, reliability, and access speeds for your own solution.

Partnering with Another Site

Partnering with another site requires joining forces with a site that sells similar products or services (or even better, complementary products). Find vendors that add value to your offerings. Use your network of contacts as a resource for finding a good match for your company. When using this approach, find a site that is heavily accessed and place your products in their storefront. You may have to expend some manpower or financial resources to develop a portion of the solution, but the effort is far less taxing than creating and hosting a solution yourself. In addition, you gain the advantage of marketing muscle as both companies devote their resources to the goal.

Joining a Mall

If you decide not to build internally, perhaps joining a mall would prove beneficial. Most malls provide services to set up your own storefront, typically within a few hours. This is a great option if you do not have the resources, technical expertise, or the money required to establish a solution internally. For smaller companies, joining a mall is a great option because it is a quick and easy method of getting your store up and running. The disadvantage is that you do not have as much flexibility in customizing your site and integrating with existing back-end systems.

In order to find the right mall for your products, search www.yahoo.com for sites that sell products that would enhance your own product sales. Select a mall with high traffic levels. Four companies dominate this market: Yahoo! (through Yahoo! Store), GeoCities (through GeoShop), iCat Mall, and iMall.

Yahoo! Store (www.yahoo.com) offers several compelling features, including a shopping-cart that allows your customers to keep track of their orders. The shopping-cart can display the sale price of products, including quantity discounts, where applicable, and offers automatic tax and shipping calculation. Once the customer selects a product, the shopping cart can even suggest related items, thus acting as a cyber-salesperson. Another important feature of Yahoo! Store is the availability of tracking tools, which let you collect and analyze certain information about your customers and their experience. For example, you can determine where your shoppers are coming from, how much they spend, how many visitors your store is receiving, sales volume, and more.

GeoCities (www.geocities.com) is best known for its free hosting services for personal home pages., The company now offers low-cost e-commerce hosting, as well, although not for free. The new service, known as "GeoShops with Transaction Processing", represents a partnership between GeoCities and Internet Commerce Services Corp., which gives small businesses the ability to securely process credit card orders in real time. GeoShops allows you to create and update your online store, process sales, interact with other store owners, and track account statistics. Store updates and management tasks can be accomplished with GeoShops' suite of file manager tools.

iCat Mall (www.icat.com) provides a user-friendly online interface for merchants to build their e-stores. Like Yahoo!Store, iCat claims that a merchant could set up shop in less than ten minutes. iCat features simple store creation and management processes, including capacity of up to 3000 items. The company allows you to put items on sale, to specify featured items, to cross-sell, and to up-sell. Their shopping cart, with several payment options, includes automatic tax and shipping charge calculations. And the site generates extensive sales, customer, and tracking reports for your analysis work.

iMALL (www.imall.com) claims to be the largest shopping mall on the Internet, with more than 1,600 hosted storefronts, and millions of visitors monthly. iMall commerce-enabled Web sites allow for secure, real-time credit card processing, merchant account services, order processing, shopping cart, product database management, store-level searching, backroom services, delivery of electronic products, and product-level searching.

When compared to the time and expense necessary to build and host your own solution, the benefits of joining a mall seem obvious. Typically, you can get started with many of these services for about $100 per month (to display about 75 items). A larger store, selling up to 1000 items costs about $300 per month. If you would like to accommodate up to 5000 items, the cost would be $700 per month. The site www.ecommercetimes.com, provides extensive reviews of online malls. It is a great place to begin your research.

Outsourcing Your Deployment

Several companies offer the total solution, from designing a Web site to actually hosting the site. This is a great option for those companies that do not have the technical expertise or resources to design, develop, and deploy a solution. The tradeoff of this option is the minimal amount of effort required, but at a possibly prohibitive cost.

How do you locate these services? The first place to look is at the sites of solution vendors discussed in the Vendor Solutions chapter, including Microsoft, IBM, BroadVision, OpenMarket, and INTERSHOP. In addition, you can conduct other research on the Internet to locate other hosting providers.

Not So Secret Success Strategies

After you deploy your site, marketing your offerings and performing updates help ensure its continued success. Develop a talented, focused team to create all of the features and functionality that set your solution apart from your competition. Create a virtual path to your Web site with creative and consistent marketing efforts. Ensure that systems are fully functional and not overloaded. Tweak and test the commerce solution to provide visitors with a new experience each visit. Study the companies that succeed in the e-commerce arena. What makes them tick? These companies use extreme creativity, focus on marketing and customer service, create a compelling user experience, and cherish the results. The inherent beauty of electronic commerce is that when you do one thing right, you get paid over and over.

12

Consortium

- How we approached e-commerce
- Skills required to implement e-commerce solutions
- Business skills versus technical skills
- How does everything fit together?

Electronic commerce is not a business driven by one individual. It includes a variety of skill sets and knowledge, both on the business side and the technology side. In order to successfully design, develop, and deploy e-commerce-based solutions, we created the Consortium of Electronic Commerce.

The Consortium of Electronic Commerce, otherwise known as the CEC, is comprised of a group of dedicated, talented professionals with the drive and commitment to mesh their business acumen with the electronic commerce expertise developed during the *E-Commerce Program*. More concisely stated, the Consortium equals business experience plus electronic commerce skills plus commitment plus communication.

Typically, the bottleneck in any e-commerce project is finding the right people to help in building the e-commerce strategy, and then developing the technical solution. With the Consortium's variety of backgrounds and skill sets, individuals with great commerce ideas can pull members from the group to build a fully functional team. These teams can then easily analyze and execute any commerce idea. What a compelling concept!

The consortium concept presented in this chapter can be used by those individuals ready to move forward and build an e-commerce site. Or use the contact information at the end of the chapter for assistance in building your solution.

A Consortium Meeting

A cool gust of wind ushers away the last hint of daylight. As we walk up the path, we turn our attention to the ornate entryway of the restaurant and its imposing guards. Tonight's gathering is invitation only. An invitation earned with four months of toil, study, and practice.

It is nine o'clock in the evening, the middle of April, and tonight we are standing at the edge of the main dining area in a trendy Dallas nightspot. Guests at Chez Taboo are all Consortium members...a growing group of professionals that have successfully completed Southern Methodist University's *E-Commerce Program*. In one corner of the restaurant, an impromptu gathering of Web designers discusses the merits of various development platforms, while a small business owner furiously scribbles out notes. In another corner, a commercial photographer commits to deliver a catalog full of product artwork in time for the recently announced unveiling of an online store's improved Web site.

Nearby, over a bowl of paella, a security expert assures two corporate executives that their recent technical implementation will prove sound. A young executive is reintroduced to an attorney, who in turn, shoos away his long-time associate, a venture capitalist. The attorney, as a member of the Consortium, has just added a new set of skills and a new perspective on the demands of business. The executive poses several questions about patents and trademarks, and their relation to electronic commerce. A database analyst joins a group of human resource professionals. She senses a bond at the table...a bond that developed as each worked their way through the commitments of the *Program*. The group seems very interested in learning all they can about the table of consultants nearby. The evening ends very late, full of smiles and handshakes.

This scene repeats itself once each month. The growing membership of the Consortium thrives on its fresh influx of members offering new approaches and talents. The opportunity to mix and mingle, to share ideas, to pool resources, and to form alliances proves very compelling.

Business Experience

The membership of the Consortium draws from many areas of the business world. Each profession lends a distinct business experience and a perspective of how that knowledge contributes to an electronic commerce solution. Representative professions fall into the categories of business principals, support services, and technical implementers.

Business Principals

The business principals group includes executives, venture capitalists, and small business owners. Members of the first group usually represent large corporations, serving as vice-presidents or MIS Directors. Small business owners comprise the third group. Bridging the two groups, often helping small business owners grow their organizations to become big business owners, are the venture capitalists.

Executives assume responsibility for leading their organizations through electronic commerce efforts. These efforts typically combine the managerial and the spiritual when organizing, delegating, financing, and encouraging team members. They ensure the successful completion of a project by harnessing the key players in the organization and clearing any roadblocks. As members of the Consortium, executives help you work with your vendors, suppliers and buyers, develop relationships to ensure customer satisfaction, and break down barriers of entry to businesses with which you would like to partner.

Steering a large corporation and driving a small business require two distinct sets of skills and temperaments. Most small businesses face more limited budgets, distribution channels, and marketing channels, as well as fewer support-ive relationships than larger organizations. Owners tend to rely on creativity, productivity, and dexterity to navigate the world of commerce. As members of the Consortium, small business owners help you expand your network, understand your ISP options, maximize your marketing budgets, and distribute goods and services efficiently.

When you have committed to a solid business plan, navi-gated all convenient financial avenues, and need to tap the resources of a seasoned veteran, the venture capitalist finds you. The resources that venture capitalists bring to the table include business acumen and financial power. As members of the Consortium, venture capitalists provide you and your solution with an experienced sounding board, access to additional funding, and a wealth of professional guidance as you continue to grow.

Support Services

In addition to the members that represent business princi-pals, an array of professionals support the primary func-tions of an electronic commerce effort. These professionals include accountants, enterprise resource planners, informa-tion technology consultants, lawyers, marketing specialists, organizational developers/transformation managers, and technical account managers.

Tax laws for selling products and services to various states within the United States and to other countries vary, and change often. Accountants play an important role when measuring the success of an electronic commerce solution. They ensure financial compliance to the dictates of govern-ment and shareholders. This is especially critical now that we are currently under a three year tax moratorium on Internet sales. As members of the Consortium, accountants guide you over the financial reporting and planning hurdles your business may face.

For companies that are currently using enterprise resource planning (ERP) systems, such as PeopleSoft or SAP, it is critical that you employ planners in the project that understand the ERP systems as well as how to integrate these components into the e-commerce system. In addition, the planners must understand the functionality of both the ERP system and the e-commerce software. This understanding ensures that you are not duplicating processes, such as inventory management, in your commerce site when SAP already provides the functionality. As members of the Consortium, enterprise resource planners assist you in integrating your e-commerce projects with your back-end systems.

A talented Information Technology (IT) consultant delivers best practices to an organization, as a product of experience with a wide variety of hardware, software and technical implementations. Typically, the consultant supports several different types of industries, such as manufacturing, retail, insurance, financial institutions, etc. The knowledge an IT consultant gains by leading companies in new technology directions can enrich your organization quickly. As members of the Consortium, IT consultants provide a wealth of solutions for you to consider.

E-commerce places many demands on the talents of lawyers by requiring an understanding of intellectual property rights, patents, trademarks, copyrights, and licensing. In addition, lawyers must understand consumer protection laws of credit card fraud, legal rights of online shoppers, and import/export laws. A significant new area of Internet law involves intellectual property and patents of new business processes that lend you a competitive advantage. In addition, since you develop brands, trademarking your company and copyrighting your information prove critical. If you are selling electronic media such as whitepapers, ensure that other vendors cannot distribute this information without paying licensing fees. As members of the Consortium, lawyers ensure your compliance in the legal arena and a competitive advantage in your marketplace.

Take a look at companies like Amazon.com or eBay. What are the barriers to entry? Anyone can sell books over the Internet or auction off a variety of goods. Consider the importance of marketing to each company's success. Marketing separates these companies from the also-rans. The strength of your marketing efforts determines the success of your solutions. Marketing specialists understand business and how to capture customers. In an e-commerce environment, marketing becomes even more critical. Your competition now lurks just a "click away" instead of twenty miles down the road. As members of the Consortium, marketing specialists contribute to your organization by studying buying trends, understanding successful advertising, capturing market demographics and geography, and developing new markets.

Reengineering your company alters both the roles and responsibilities of your staff. And if you attempt to automate the business processes between two or more companies, the changes become even more complex. The role of the organizational developer/transformation manager (OD/TM) involves surveying the overall organization and determining how the transformation process affects various groups. Once they have categorized the groups, the OD/TMs develop a plan of action to help transition people into their new roles. These plans may include specialized training, informational sessions, and some hand holding. An OD/TM also determines (with the assistance of the company) how to involve employees in the decision-making process. This sense of ownership reduces any resistance to change. As members of the Consortium, the organizational developer/transformation manager consults with your company to ease your transition to the world of electronic commerce.

Since an e-commerce solution requires a vast amount of technical knowledge on a variety of software and hardware platforms, technical support is crucial. A technical account manager quickly resolves technical problems that occur during the development and deployment of the E-commerce solution. As members of the Consortium, technical account managers help you develop your internal support structure for both technical and customer support.

Technical Implementers

In addition to the members that represent business principals or support the primary functions of an e-commerce effort, the Consortium includes technical implementers. Members of this category design, develop, and deploy the hardware and software that comprises an e-commerce solution. These professionals include credit card system integrators, database analysts, interface designers, photographers, programmers, security consultants, and Web developers.

A major component of an e-commerce deployment is the credit card verification system. Integrators of these systems possess the knowledge of setting up the entire process, from working with financial institutions and credit card companies to integrating the software into the e-commerce software package. In addition, they understand how product returns, currencies, and encryption methods affect your commerce site. Setting up and integrating a credit card system, practiced so infrequently, dictates outsourcing, rather than using your resources to train someone internally. Getting paid for transactions is how you ultimately judge your efforts. As members of the Consortium, credit card system integrators help bridge the gap between transactions and payments.

Most commerce solutions require integration into a relational database. A database analyst (DBA) assumes responsibility for this integration. Skills include expertise in relational database security, database structures, server optimization, and other database components. A DBA focuses on the capabilities of the e-commerce software and connectivity tools required to tie into these systems. As members of the Consortium, database analysts perform the tasks necessary to link your back-end systems to your electronic commerce implementation.

You have been reminded not to judge a book by its cover. But your customers judge your Web site by its "cover." The only thing your customers see when accessing your Web site is the interface (the colors, graphics, text, and layout). An interface designer ensures that your Web site appeals to the eyes (and maybe ears) and proves easy to navigate. Sites that incorporate glaring or cryptic interfaces are soon abandoned. Someone out there offers a compelling site. As members of the Consortium, interface designers ensure that the compelling site belongs to you.

If you want to display your products online, taking the right picture can make or break a product sale. Professional photography pays for itself many times over. For example, if you are selling jewelry, the photographer captures the true beauty of a piece. Otherwise, beauty is hidden within a two-dimensional presentation. Remember that "perception is nine-tenths of the law." As members of the Consortium, photographers provide the raw material that drives your marketing efforts.

Since e-commerce involves integrating software and designing workflow processes, programmers can support your design efforts. Programming skills include languages such as Java and Visual Basic, and technologies such as JavaScript, DHTML, HTML, CGI, and cookies. As members of the Consortium, programmers assist site design, develop custom applications, and debug uncooperative code.

Hackers reside all over the world and within your organization. Adequate security ensures that your company and customer information remains separated from the eyes of the public. A security consultant understands where you are vulnerable and how to shore up your defense. As members of the Consortium, security consultants assume responsibility for the safety of your data and the continuity of your site.

A Web site developer understands the technology required to implement an electronic commerce project, understands the transition from a brick-and-mortar economy to an electron-based economy, and has the skills to create a technical solution. This category complements the interface designer in that the developer devises the infrastructure, while the designer creates the user interface. As members of the Consortium, Web site developers analyze the range of options, design a system to satisfy your specific objectives, and minimize your expenses.

The combination of business principals, support professionals and technical implementers among the membership of the Consortium ensures that your requirements for business knowledge and technical skills are addressed.

Electronic Commerce Skills

Each Consortium member graduates the *E-Commerce Program*...a program that emphasizes a combination of architecture theory and hands-on implementation. Upon graduation, participants demonstrate their ability to architect an e-commerce solution from defining business goals and strategies to understanding the technologies needed to implement the solution. In addition, students become familiar with Lotus, IBM, Microsoft, and Netscape e-commerce solutions by reviewing back-end servers, Web commerce servers, connectivity tools, and front-end architecture. The level of instruction that the graduates now have in common allows them to work more efficiently and effectively.

The initial module introduces students to e-commerce business solutions, e-commerce technical architecture, back-end connectivity, Web servers, commerce servers, front-end clients, credit card verification systems, and firewalls.

The second module provides students with the knowledge necessary to architect an e-commerce solution using Microsoft's Site Server. Features including the IIS architecture, security, domain model, and Microsoft's Site Server replication, shopping and merchandising features, personalization and membership, Ad Server and Site Server Analysis are used as a basis for the design.

The third module provides students with the knowledge required to architect an e-commerce solution using Domino.Merchant and IBM's Net.Commerce. Features such as the Domino architecture, security, domain model, replication and Domino.Merchant's Site Creator, Content Manager, shopping and merchandising features, CyberCash payment system and TAXWARE International are used as a basis for the design.

The fourth module provides students with an understanding of the tools that allow back-end connectivity to relational databases, data warehouses and enterprise resource planning applications.

The fifth module introduces students to several front-end tools used to enhance the commerce server applications.

These tools include Java applications, push-pull technologies and other front-end plug-ins.

The final module provides students with an understanding of server security in an intranet/extranet architecture. These features are demonstrated by installing and configuring several types of firewall and proxy server products. Additional topics covered include SSL security, encryption, operating system security, smart cards, digital signatures, electronic wallets, and Java security.

The *E-Commerce Program* offers an exhaustive, hands-on, vendor-neutral program of study that allows students to excel in the world of electronic commerce. Consortium members harness the skills developed within the *Program* and prior business acumen to improve the performance of any commercial ventures they undertake.

Commitment

In addition to their common base of e-commerce knowledge, the members of the Consortium are also bound by an exceptional level of commitment. The *E-Commerce Program* requires a time commitment, a financial commitment, and a commitment of resources.

The six courses that comprise the *E-Commerce Program* require a commitment of 192 contact hours over a period of four months. Keep in mind that a majority of the students work full-time, and attend the class sessions on their own time in the evening. Although the pace challenges, students are buoyed by the sense of accomplishment generated by each session.

Each student also commits resources to the success of the *Program*. Each student brings a different professional background to the classroom environment. A student's willingness to share that experience and to support the work of fellow students naturally leads to the enrichment of the Consortium concept.

Finally, the *Program* requires a financial commitment. The breadth and depth of the instruction, in addition to the vast array of software products introduced and explored, assume a expense on par with other professional programs around the country. However, the skills that develop enhance a graduates employment opportunities and the magnitude of their success in e-commerce ventures.

How Members Are Connected

Members of the Consortium communicate with each other in a variety of forums. The specific means of requesting information, distributing knowledge, sharing leads, and introducing new members varies among monthly gatherings, e-mail messages, a bulletin board posting, and establishing a virtual office.

The monthly gatherings, as described at the outset of the chapter, provide an opportunity to meet new members, to share ideas, to pool resources, and to form project-based alliances. Social gatherings are beneficial, allowing face-to-face communication. They add a level of trust and security.

The Consortium sponsors a bulletin board that allows members to post queries, solicit participation, answer questions, and air opinions. Frequently, bulletin board postings assist the creative process of forming teams. One student explains that the function of the Consortium of Electronic Commerce bulletin board is "to provide a forum for members who are undertaking an endeavor or to float ideas and find talent that they would not otherwise have access to. For instance I might try to find somebody with an art background to be my partner in a project, since design is not my background, by posting to the board."

E-mail allows Consortium members to communicate one to one or one to many. This more targeted means of communication, in comparison with bulletin board postings, allows members to confer when topics require more discretion or are not of interest to the general community.

When members join forces to tackle a specific project, communication takes the form of a virtual office. The project team consists of both the customer and Consortium members participating in the engagement. In a virtual office, a project database, hosted at either an ISP or at the client site, allows members of the project team to view and post information for the duration of the project. If a team member is not connected via a LAN, the server is set up to allow dial-up access or set up with an Internet Service Provider (ISP). Typically the project databases grow to include project management documentation, infrastructure standards and application standards.

The project management databases contain most of the project information, including project plans, action items, project deliverables, meeting minutes, team member profiles, and miscellaneous documentation. The project plan section tracks project scope, project deliverables, timelines, and resources for the project. As you develop a solution, an action item table tracks the status of each action item that arises in meetings and in general discussions. Project deliverables are descriptions of the business process review, application design specs, infrastructure design, workflow diagrams, and other documentation eventually delivered to the customer. Meeting minutes capture the detail of each project meeting or conference call. Posting meeting minutes to a central area helps to ensure that everyone on the team is aware of decisions made during meetings and the resulting assigned action items. The member profile section contains the name, phone number, e-mail address, availability, and skills of each team member. Miscellaneous documentation explains products used or reviewed for the e-commerce solution, as well as hardware spec sheets and other resource material.

Another database contains the infrastructure guidebook. This guidebook details the entire infrastructure of the organization, including operating system configuration, server sizing and placement, communication links between sites, security, replication schedules, and naming standards for servers and domains.

A third server contains the application guidebook. The application guidebook specifically documents the target application. This detail includes application design, application integration, form and field level design, process code, application security and interface design. One of the things that is extremely valuable about having all of this information online is the immediacy of access. In addition, all team members can add, update, and delete information at one convenient location.

Going It Alone?

When you are finally ready to design, develop, and deploy an electronic commerce solution, you can select among three groups of professionals to perform the work. The first group is comprised of members of your own staff. The second group is comprised of consultants brought in for the initial phases of your project. The third group is comprised of Consortium professionals, partnering with your company for the duration of the deployment. Each of these three groups is worthy of consideration.

If you decide to use your company's current staff to create an e-commerce solution, consider the following four issues. One, can you steal employees from your IT group and from your marketing group or do you want them to work on the projects you hired them for originally? If they find the time and resources to work on your project, what falls behind? Two, what do you do if your project is a raving success? Do you reassign these staff members permanently as e-commerce only. Three, how do you address the maintenance requirements or ramp-up demands that lurk around the corner? Now that you are successfully deployed, do you need someone else to maintain or grow the system? Four, is this really your area of expertise? Or are you just looking for a more effective way to be more effective?

Right-Click to View Source

If you decide to contract with a consultant group to create an e-commerce solution, consider the following four issues. One, determine whether your consultants are vendor-neutral. If you are a hammer, every problem looks like a nail. Can they suggest and deliver a solution from the vendor that most effectively addresses your concerns? Two, is your solution destined to become someone else's solution tomorrow or for that matter, was it someone else's solution yesterday? Do not fund your competitor's solution. Three, how long will the consultants remain at your location? Just as you face when using your own staff members, the issue of maintenance and ramp-up must be addressed. Four, what price will you pay for success?

Figure 12.1 CEC Home Page

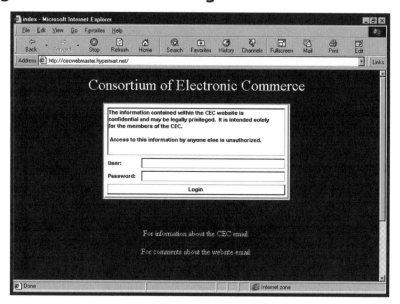

A simple exercise suggests the type of competition that is prevalent within the development environment of electronic commerce. Take a look at any Web page. Consider the effort and skill that is reflected in its design. Now, right-click on the page to display its source code. A single click reveals the underpinnings of the interface design. Compare the Web page in Figure 12.1 to the source code displayed in Figure 12.2. This source code represents only a small part of the total development effort. However, be sure to consider any area where your organization and its efforts may be vulnerable. Harness the expertise of your employees and your consultants to create a strong, secure solution.

Figure 12.2 Source Code of the Home Page

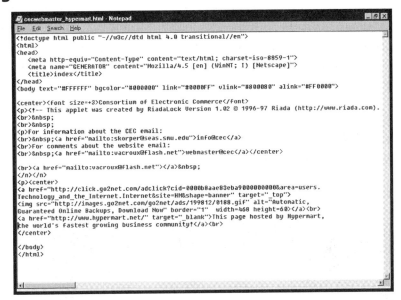

Partnering with the Consortium

If you decide to partner with the Consortium, consider the following issues. One, what convinces you that you made the correct decision? Two, how do you address maintenance requirements or ramp-up demands? Three, do your candidate solutions come from a wide pool of vendors? Four, how will your new team members interact with your current team members?

Turn on your television. Open a magazine. Everyone is suddenly an e-commerce player. Opportunities in this area continue to multiply, so everyone labels themselves an expert. How do you tilt the odds in your favor? When you join forces with members of the Consortium, you draw on the unique talents and diverse backgrounds of specific individuals, as your particular project demands. Instead of hiring an entire group of consultants, the Consortium concept allows you to partner with professionals, judged on their own merits. No more hiring the whole team, no more firing the whole team. The diversity of the Consortium membership ensures that your project incorporates fresh ideas and tested approaches.

The soundest model for developing e-commerce solutions combines implementation driven by an IT group with design and development driven by a business group. The business group understands your marketplace, your customers, your partners, and your challenges. Let them lead the charge. When your now profitable solution requires maintenance or some form of scalability (the capacity to ramp up or down), draw on the characteristic of the Consortium concept that allows members to float between projects or to devote themselves to a project exclusively.

If all the pieces fall into place, except for the budget, consider an alternative. Buy either the solution or the transactions. Either pay for the entire solution or use a percentage of the profit from transactions on your Web site to fund its development. A profitable solution makes everyone happy.

Remember that the Consortium is the smallest and the largest source of experts. The compelling advantage is that you can draw on as many and as much of the resources as your solution requires. What would you have to do as a company to build your own consortium?

How Does It Work?

Many types of organizations take full advantage of their relationship with the Consortium. To learn how these our professionals contribute to a variety of organizations, take a look at a big business, Texas Utilities Company, a small business, Exceptional Products, Inc., and a non-profit group, the Volunteer Center of Dallas.

TXU

TXU, formerly known as Texas Utilities Company, based in Dallas, Texas, is an investor-owned company providing electric and natural gas services, energy marketing, telecommunications, and energy-related services domestically and internationally. TXU is one of the largest investor-owned energy service companies, with assets over $40 billion, serving approximately nine million customers worldwide. More information about TXU and their many services is available at www.txu.com.

The latest TXU challenge involves harnessing the power of electronic commerce to improve the competitive capabilities of their multifaceted utility. The professionals of TXU addressed this challenge by focusing on adding value to the daily lives of their residential and business customers with creative e-commerce solutions.

Within their Web site, TXU provides a variety of value-added services to residential customers. Home Energy Audit generates a free online profile of a customer's energy use, including tips on managing energy more efficiently. HEET (Home Energy Efficiency Tour) is a virtual tour of a home that teaches techniques for more efficient heating and cooling, insulation, and lighting. TXU's Web site also features E-Choice, a system that provides expert advice about replacing a heating or air conditioning system and referrals, if requested. In order to ask questions or express concerns, residents with access to the TXU Web site use e-mail to contact energy experts. These features represent only a few of the many ways that TXU adds value to the lives of their residential customers with electronic commerce.

TXU also supports the performance of their commercial customers, especially in the area of improved billing practices. By implementing electronic data interchange (EDI), the computer-to-computer exchange of documents in standardized, electronic formats, TXU offers paperless billing. This method of billing saves between $1.30 and $5.50 per bill by eliminating the time and labor required to process and pay bills manually. EDI represents a business strategy that deploys a technical solution to improve business relationships and to meet target objectives. It also forms the backbone of many business-to-business solutions.

TXU's relationship with the Consortium, although informal, pays dividends. Employees of TXU, whether in management, engineering, or even human resources, participate in the *E-Commerce Program*. By gaining an understanding of how to capitalize on the potential of electronic commerce, managers are able to create compelling value-added services for their customers. Engineers, who learn techniques for implementing e-commerce solutions, contribute their expertise to the deployment process and solve technical problems on the job. Human resource professionals who now understand the concepts and vocabulary associated with e-commerce communicate much more effectively when screening potential employees. In addition, these individuals realize the benefits of Consortium membership as a source of creative solutions, technical support, and a solid education.

Exceptional Products, Inc.

Small business principals also take advantage of membership in the Consortium. Elliott Brackett, a principal of Exceptional Products, Inc. (EPI), recently graduated from the *E-Commerce Program: Architecting the Solution*, and put his knowledge and experience to work deploying a successful e-commerce solution. EPI, a direct response marketing company, specializes in creating nationally known brand names for new products, by using infomercials, direct response television (DRTV) and other direct marketing methods. As a integral part of its presentation, EPI finances multi-million dollar marketing campaigns for winning products, protects products in a competitive marketplace, determines effective price points and leverages knowledge.

EPI faces two significant challenges to its business model. The first is to find the great products. Their Web site (www.sellontv.com) addresses this challenge by soliciting new products. The company provides inventors of potentially compelling products with a customized campaign to sell their products, as described below.

The EPI Web site instructs its audience to "simply pick up the telephone and call us or send an e-mail to tell us about your product. After a careful and timely review (usually within 5 business days), we'll respond in writing whether we think EPI can help." In addition, EPI reassures its audience that "all our agreements are performance related, so there is absolutely no risk on your part. We'll create, test, and evaluate an original DRTV campaign featuring your product...typically within 120 days. If the market responds to your product the way it has to other successful products we've launched, we'll become your turnkey solution via television, radio, print, and retail. If our campaign doesn't produce immediate results, you've lost nothing and still gained additional insight into your product."

By presenting this sales pitch on their Web site, EPI can broadcast its message very efficiently and inexpensively. In addition, their customers can immediately ask questions or solicit more information via e-mail.

The second challenge for EPI is that some products have to be explained to be sold. For example, HobbyLab's SR-71 Black Bird, a relatively expensive and sophisticated model jet for hobbyists, sold above expectations once its story was presented. Emphasizing its functionality and quality generated interest in the model...interest that sometimes does not exist when products are distributed only via store shelves.

By combining direct response television with the Web site www.soldontv.com, EPI creates a virtual storefront on the Internet. Products featured at this storefront have traditionally been sold only through television, and only as impulse purchases. By using a Web site to market these items, EPI enhances the exposure of the products, reduces the cost of marketing, improves customer support, and allows consumers to interact directly with the developers and distributors. Each of these features enriches a customer's shopping experience, and thereby improves the sales results.

The Volunteer Center of Dallas

The Consortium also contributes to the community by supporting non-profit organizations. The Volunteer Center of Dallas (VCD) serves as a recent example of the success achievable when necessity outweighs funding. The following profile is derived from a report researched and written by John Eaton.

The mission of the Volunteer Center of Dallas is to promote volunteering and to refer human and material resources to nonprofit organizations. For 28 years the VCD has supplied countless volunteers and many millions of dollars in material goods to the nonprofit agencies of our community. In Dallas County, the VCD serves more than 1,100 nonprofit agencies with an extensive variety of programs and services. Last year, the VCD made more than 91,000 volunteer referrals and placed more than $8 million in goods with nonprofit agencies.

The VCD wanted to improve their services by adding an online bookstore to their current list of offerings. The audience for this site includes people who use VCD agency services, VCD members, and the public at large.

The resulting e-commerce solution, called *A New Chapter*, enables volunteers in service to non-profit organizations, and its clients to make online purchases of self-help and educational books. The Web site offers a list of subjects and titles that the customer can select, add to a shopping basket, and then pay for via a secure payment site.

The business process involved in moving books from donation to sale remained fairly uncomplicated. The VCD draws merchandise from close-out merchandise buyers (COMBs), book publishers, overstock jobbers, self-publishers and other second-hand book stores. Companies that support the VCD use the "We Support the Volunteer Center" logo and slogan on their letterheads and Web sites. Each logo represents a clickthrough to the VCD homepage describing the VCD and its mission. Banner ads, with clickthrough capabilities, are placed anywhere charities, COMBs, book publishers, overstock jobbers, self-publishers, and other second-hand book stores are found on the Net. Books are received by the VCD and delivered to the warehouse. Titles are indexed by subject using MS FrontPage Site Manager. When a customer purchases a book, the inventory record stored on the back-end database decreases. The order is pulled from the warehouse and then shipped. The goal, in alignment with the business process, was to create a fully integrated site that allows books contributed to VCD to be tracked from the point of contribution to the point of retail sale.

John Eaton, Mark Hanson, and Bill Kirkland, members of the Consortium as well as employees of E3 Solutions, designed, developed, and deployed the solution. The implementation seamlessly integrated a FrontPage solution with their current Microsoft-based legacy system. Eaton, as content manager and site architect, developed and defined the indexes by subject matter. He also architected the workflow of the Web site. Hanson, system configuration and programming analyst, configured the Web server and data servers. His extensive knowledge in programming and installation helped speed up the delicate configuration process. Kirkland, Web designer and beta tester, helped create the look of the site, making it attractive and easy to read. He then tested the site for ease of use and accuracy. Changes and improvements were based on his thorough critique.

For More Information...

What would you have to do as a company to replicate this organization? The Consortium of Electronic Commerce offers some compelling benefits to its corporate partners. Can the Consortium help you reach your company's goals? If it sounds like we can help you, and you want to discuss the Consortium and its services in more detail, please write us:

Consortium of Electronic Commerce
c/o Steffano Korper
10801 Branch Oaks Circle
Dallas, Texas 75230

or contact us by e-mail at skorper@seas.smu.edu

Authors' Notes

This book is a direct result of over 10 years of industry experience and education. Our goal is to provide you with the information and knowledge necessary to start moving your companies forward in the area of electronic commerce. Our background of industry and educational experience helps ensure your success.

In 1994, we developed a series of courses called the *Networking Technologies Program* at Southern Methodist University (SMU). First of its kind, the program focused on Internet technologies prior to the now common industry hype. The goal was to create a program that provided "real life" practice. The strength of the program was its focus on network integration. By not focusing on a single platform, the *Networking Technologies Program* taught students to integrate entire network systems. Students with little prior computer knowledge felt confident that they understood the technologies, had the ability to implement these technologies, and completely changed careers in the field of information technology.

The *Networking Technologies Program*, a huge success, generates enthusiasm among its graduates. One student said that "words cannot express what I have learned from you these past three months. You opened my eyes to a world of networking during the Info Session. Though unemployed, I made the commitment to attend the *Networking Technologies* class at SMU's School of Engineering and Applied Science. You introduced me to the world of the Internet and its many navigation tools. It is evident that you knew the software and other systems, so I tried to cling to every spoken word." Another reinforced this expression by describing his recent job interview…"he [the recruiter] also asked about the *Networking Technologies Program* that I am taking and was very impressed with the SMU-backed certification. He said he would consider this certification more valuable than a CNE."

In addition to these accolades, the tangible results, in one case, included the hiring of an entire *Networking Technologies Program* class…en masse.

After creating the program at SMU, we returned to work in the computer industry. Steffano enhanced his industry experience as a director of MIS for Wyndham Hotels and Resorts. He is currently vice-president of E-commerce Solutions at Going Beyond E-commerce Technologies LLC, working closely with corporations in developing e-commerce strategies, reviewing competitor markets, and developing systems that exceed typical e-commerce efforts.

Juanita joined Lotus Consulting, assuming the lead technical role in the design of total systems solutions for large engagements. She deployed solutions within aircraft manufacturing, telecommunications, research, government, and "Big 6" organizations, with an emphasis on process reengineering, system-wide integration, and business-to-business commerce. These deployments included responsibility for designing and implementing enterprise-wide applications that integrated with enterprise resource planning systems, Internet technologies, and relational and transaction-based systems. Juanita relished her experience as a top technical architect.

Then, as we mentioned earlier, after seeing the Forrester prediction for the Internet's trade potential, we both changed directions and focused on electronic commerce. This led to the creation of the *E-Commerce Program*.

Index

ISBN 0-12-421160-7